Marx Joyce
Hardy Emerson Austen
Defoe Abbott Montaigne Chesterton Cooper Hugo
Melville Machiavelli Eliot
Stoker Carroll Christie Haggard Molière Grimm
Wilde Maupassant Byron Schiller
Garnett Fitzgerald Engels
Goethe Einstein Hawthorne Smith Kafka
Cotton Dostoyevsky Hall
Baum Henry Kipling Doyle Willis
Leslie Dumas Flaubert Turgenev Nietzsche Balzac
Stockton Vatsyayana Crane
Burroughs Verne
Curtis Tocqueville Whitman Gogol Vinci
Homer Widger Tolstoy Busch
Darwin Thoreau Twain
Potter Freud Zola Lawrence Scott Harte
Kant Jowett Stevenson Dickens Plato
Andersen Burton Hesse
London Descartes Cervantes
Poe Aristotle Wells Voltaire Cooke
Hale James Hastings
Bunner Shakespeare Irving
Richter Chambers da Vinci
Doré Dante Chekhov Shaw Benedict Alcott
Swift Wodehouse Pushkin Newton

# ⑪ tredition®

tredition was established in 2006 by Sandra Latusseck and Soenke Schulz. Based in Hamburg, Germany, tredition offers publishing solutions to authors and publishing houses, combined with world-wide distribution of printed and digital book content. tredition is uniquely positioned to enable authors and publishing houses to create books on their own terms and without conventional manu-facturing risks.

For more information please visit: www.tredition.com

## TREDITION CLASSICS

This book is part of the TREDITION CLASSICS series. The creators of this series are united by passion for literature and driven by the intention of making all public domain books available in printed format again - worldwide. Most TREDITION CLASSICS titles have been out of print and off the bookstore shelves for decades. At tredi-tion we believe that a great book never goes out of style and that its value is eternal. Several mostly non-profit literature projects pro-vide content to tredition. To support their good work, tredition donates a portion of the proceeds from each sold copy. As a reader of a TREDITION CLASSICS book, you support our mission to save many of the amazing works of world literature from oblivion. See all available books at www.tredition.com.

##  Project Gutenberg

The content for this book has been graciously provided by Project Gutenberg. Project Gutenberg is a non-profit organization founded by Michael Hart in 1971 at the University of Illinois. The mission of Project Gutenberg is simple: To encourage the creation and distribu-tion of eBooks. Project Gutenberg is the first and largest collection of public domain eBooks.

# Farm drainage The Principles, Processes, and Effects of Draining Land with Stones, Wood, Plows, and Open Ditches, and Especially with Tiles

Henry F. (Henry Flagg) French

# Imprint

This book is part of TREDITION CLASSICS

Author: Henry F. (Henry Flagg) French
Cover design: Buchgut, Berlin – Germany

Publisher: tredition GmbH, Hamburg - Germany
ISBN: 978-3-8472-3056-4

www.tredition.com
www.tredition.de

"Read, not to contradict and to confute, nor to believe and take for granted, but to weigh and consider." — Bacon.

"The first Farmer was the first man, and all nobility rests on the possession and use of land." — Emerson.

to
The Honorable Simon Brown,
of Massachusetts,
A Lover of Agriculture, and a Progressive Farmer,
whose Words and Works are so well devoted to Improve the Condition
of Those who Cultivate the Earth,
this Book is Inscribed, as a Testimonial of Respect and Personal Esteem,
by his Friend and Brother,

The Author.

# PREFACE.

The Agriculture of America has seemed to me to demand some light upon the subject of Drainage; some work, which, with an exposition of the various theories, should give the simplest details of the practice, of draining land. This treatise is an attempt to answer that demand, and to give to the farmers of our country, at the same time, enough of scientific principles to satisfy intelligent inquiry, and plain and full directions for executing work in the field, according to the best known rules. It has been my endeavor to show what lands in America require drainage, and how to drain them best, at least expense; to explain how the theories and the practice of the Old World require modification for the cheaper lands, the dearer labor, and the various climate of the New; and, finally, to suggest how, through improved implements and processes, the inventive genius of our country may make the brain assist and relieve the labor of the hand.

With some hope that my humble labors, in a field so broad, may not have entirely failed of their object, this work is offered to the attention of American farmers.

H. F. F.

The Pines, Exeter, N. H., March, 1859.

# LIST OF ENGRAVINGS.

- Pratt's Tile Machine
- Tiles, laid well and ill
- Square and Plumb-Level
- Spirit Level
- Staff and Target
- Span, or A Level
- Grading Trenches by Lines
- Challoner's Level
- Drain Spades
- Spade with Spur
- Common Shovel and Spade
- Long-handled Round Shovel
- Shovel Scoop
- Irish Spade
- Birmingham Spades
- Narrow Spades
- English Bottoming Tools
- Drawing and Pushing Scoops
- Pipe-Layer
- Pipe-Laying
- Pick-axes
- Drain Gauge
- Elkington's Auger
- Fowler's Drain Plow
- Pratt's Ditcher
- Paul's Ditcher
- Germination
- Land before Drainage and After
- Heat in Wet Land
- Cracking of Clays
- Drainage of Cellar
- Drainage of Barn Cellar
- Plan of Rand's Drainage
- Plan of H. F. French's Drainage

# CONTENTS.

Subsoil plow, and Frost.—Effect of Frost on Tiles and Aqueducts.

Drainage Improves the Quality of Crops.—Drainage prevents Drought.—Drained Soils hold most Water.—Allow Roots to go Deep.—Various Facts.

- CHAPTER XV.

- TEMPERATURE AS AFFECTED BY DRAINAGE.

- Drainage Warms the Soil in Spring.—Heat cannot go down in Wet Land.—Drainage causes greater Deposit of Dew in Summer.—Dew warms Plants in Night, Cools them in the Morning Sun.—Drainage varies Temperature by Lessening Evaporation.—What is Evaporation.—How it produces Cold.—Drained Land Freezes Deepest, but Thaws Soonest, and the Reasons.

- CHAPTER XVI. [xi]

- POWER OF SOILS TO ABSORB AND RETAIN MOISTURE.

- Why does not Drainage make the Land too Dry?—Adhesive Attraction.—The Finest Soils exert most Attraction.—How much Water different Soils hold by Attraction.—Capillary Attraction, illustrated.—Power to Imbibe Moisture from the Air.—Weight Absorbed by 1,000 lbs. in 12 Hours.—Dew, Cause of.—Dew Point.—Cause of Frost.—Why Covering Plants Protects from Frost.—Dew Imparts Warmth.—Idea that the Moon Promotes Putrefaction.—Quantity of Dew.

- CHAPTER XVII.

- INJURY OF LAND BY DRAINAGE.

- Most Land cannot be Over-drained.—Nature a Deep drainer.—Over-draining of Peaty Soils.—Lincolnshire Fens. Visit to them in 1857.—56 Bushels of Wheat to the Acre.—Wet Meadows Subside by Drainage.—Conclusions.

- CHAPTER XVIII.

- OBSTRUCTION OF DRAINS.

- Tiles will fill up, unless well laid.—Obstruction by Sand or Silt.—Obstructions at the Outlet from Frogs, Moles, Action of Frost, and Cattle.—Obstruction by Roots.—Willow, Ash, &c., Trees capricious.—Roots enter Perennial Streams.—Obstruction by Mangold Wurtzel.—Obstruction by Per-Oxide of Iron.—How Prevented.—Obstructions by the

# FARM DRAINAGE. [13]

# CHAPTER I.
# INTRODUCTORY.

Why this Treatise does not contain all Knowledge.—Attention of Scientific Men attracted to Drainage.—Lieutenant Maury's Suggestions.—Ralph Waldo Emerson's Views.—Opinions of J. H. Klippart, Esq.; of Professor Mapes; B. P. Johnston, Esq.; Governor Wright, Mr. Custis, &c.—Prejudice against what is English.—Acknowledgements to our Friends at Home and Abroad.—The Wants of our Farmers.

A Book upon Farm Drainage! What can a person find on such a subject to write a book about? A friend suggests, that in order to treat any one subject fully, it is necessary to know everything and speak of everything, because all knowledge is in some measure connected.

With an earnest endeavor to clip the wings of imagination, and to keep not only on the earth, but to burrow, like a mole or a sub-soiler, *in* it, with a painful apprehension lest some technical term in Chemistry or Philosophy should falsely indicate that we make pretensions to the character of a scientific farmer, or some old phrase of law-Latin should betray that we know something besides agriculture, and so, are not worthy of the confidence of practical men, we have, nevertheless, by some means, got together more than a bookfull of matter upon our subject. [14]

Our publisher says our book must be so large, and no larger—and we all know that an author is but as a grasshopper in the hands of his publisher, and ought to be very thankful to be allowed to publish his book at all. So we have only to say, that if there is any chapter in this book not sufficiently elaborate, or any subject akin to that of drainage, that ought to have been embraced in our plan and is not, it is because we have not space for further expansion. The reader has our heartfelt sympathy, if it should happen that the very topic which most interests him, is entirely omitted, or imperfectly treated; and we can only advise him to write a book himself, by way of showing proper resentment, and put into it everything that everybody desires most to know.

A book that shall contain all that we do *not* know on the subject of drainage, would be a valuable acquisition to agricultural literature, and we bespeak an early copy of it when published.

Irrigation is a subject closely connected with drainage, and, although it would require a volume of equal size with this to lay it properly before the American public, who know so little of water-meadows and liquid-manuring, and even of the artificial application of water to land in any way, we feel called upon for an apology for its omission.

Lieutenant Maury, whose name does honor to his nation over all the civilized world, and on whom the blessings of every navigator upon the great waters, are constantly showered, in a letter which we had the honor recently to receive from him, thus speaks of this subject:

"I was writing to a friend some months ago upon the subject of drainage in this country, and I am pleased to infer from your letter, that our opinions are somewhat similar. The climate of England is much more moist than this, though the amount of rain in many parts of this country, is much greater than the amount of rain there. It drizzles [15] there more than it does here. Owing to the high dew point in England, but a small portion only — that is, comparatively small — of the rain that falls can be evaporated again; consequently, it remains in the soil until it is drained off. Here, on the other hand, the clouds pour it down, and the sun sucks it up right away, so that the perfection of drainage for this country would be the very reverse, almost, of the drainage in England. If, instead of leading the water off into the water-veins and streams of the country, as is there done, we could collect it in pools on the farm, so as to be used in time of drought for irrigation, then your system of drainage would be worth untold wealth. Of course, in low grounds, and all places where the atmosphere does not afford sufficient drainage by evaporation, the English plan will do very well, and much good may be done by a treatise which shall enable owners to reclaim or improve such places."

Indeed, the importance of this subject of drainage, seems all at once to have found universal acknowledgement throughout our country, not only from agriculturists, but from philosophers and men of general science.

Emerson, whose eagle glance, piercing beyond the sight of other men, recognizes in so-called accidental heroes the "Representative men" of the ages, and in what to others seem but caprices and conventionalisms, the "Traits" of a nation, yet never overlooks the practical and every-day wants of man, in a recent address at Concord, Mass., the place of his residence, thus characteristically alludes to our subject:

"Concord is one of the oldest towns in the country — far on now in its third century. The Select-men have once in five years perambulated its bounds, and yet, in this year, a very large quantity of land has been discovered and added to the agricultural land, and without a murmur of complaint from any neighbor. By drainage, [16] we have gone to the subsoil, and we have a Concord under Concord, a Middlesex under Middlesex, and a basement-story of Massachusetts more valuable than all the superstructure. Tiles are political economists. They are so many Young-Americans announcing a better era, and a day of fat things."

John H. Klippart, Esq., the learned Secretary of the Ohio Board of Agriculture, expresses his opinion upon the importance of our subject in his own State, in this emphatic language:

"The agriculture of Ohio can make no farther marked progress until a good system of under-drainage has been adopted."

A writer in the *Country Gentleman*, from Ashtabula County, Ohio, says: — "One of two things must be done by us here. Clay predominates in our soil, and we must under-drain our land, or sell and move west."

Professor Mapes, of New York, under date of January 17, 1859, says of under-draining:

"I do not believe that farming can be pursued with full profit without it. It would seem to be no longer a question. The experience of England, in the absence of all other proof, would be sufficient to show that capital may be invested more safely in under-draining, than in any other way; for, after the expenditure of many millions by English farmers in this way, it has been clearly proved that their increased profit, arising from this cause alone, is sufficient to pay the total expense in full, with interest, within twenty years, thus leaving their farms increased permanently to the amount of the total cost, while the income is augmented in a still greater ratio. It is quite doubtful whether England could at this time sustain her increased population, if it were not for her system of thorough-drainage. In my own practice, the result has been such as to convince me of its advantages, and I [17] should be unwilling to enter into any new cultivation without thorough drainage."

B. P. Johnson, Secretary of the New York Board of Agriculture, in answer to some inquiries upon the subject of drainage with tiles, writes us, under date of December, 1858, as follows:

"I have given much time and attention to the subject of drainage, having deemed it all-important to the improvement of the farms of our State. I am well satisfied, from a careful examination in England, as well as from my observation in this country, that tiles are far preferable to any other material that I know of for drains, and this is the opinion of all those who have engaged extensively in the work in this State, so far as I have information. It is gratifying to be assured, that during the year past, there has been probably more land-draining than during any previous year, showing the deep interest which is taken in this all-important work, so indispensable to the success of the farmer."

It is ascertained, by inquiry at the Land Office, that more than 52,000,000 acres of swamp and overflowed lands have been selected under the Acts of March 2d, 1849, and September 28th, 1850, from the dates of those grants to September, 1856; and it is estimated that, when the grants shall have been entirely adjusted, they will amount to 60,000,000 acres.

Grants of these lands have been made by Congress, from the public domain, gratuitously, to the States in which they lie, upon the idea that they were not only worthless to the Government, but dangerous to the health of the neighboring inhabitants, with the hope that the State governments might take measures to reclaim them for cultivation, or, at least, render them harmless, by the removal of their surplus water.

Governor Wright, of Indiana, in a public address, [18] estimated the marshy lands of that State at 3,000,000 acres. "These lands," he says, "were generally avoided by early settlers, as being comparatively worthless; but, when drained, they become eminently fertile." He further says: "I know a farm of 160 acres, which was sold five years ago for $500, that by an expenditure of less than $200, in draining and ditching, has been so improved, that the owner has refused for it an offer of $3,000."

At the meeting of the United States Agricultural Society, at Washington, in January, 1857, Mr. G. W. P. Custis spoke in connection with the great importance of this subject, of the vast quantity of soil—the richest conceivable—now lying waste, to the extent of 100,000 acres, along the banks of the Lower Potomac, and which he denominates by the old Virginia title of *pocoson*. The fertility of this reclaimable swamp he reports to be astonishing; and he has corroborated the opinion by experiments which confounded every beholder. "These lands on our time-honored river," he says, "if brought into use, would supply provisions at half the present cost, and would in other respects prove of the greatest advantage."

The drainage of highways and walks, was noted as a topic kindred to our subject, although belonging more properly perhaps, to the drainage of towns and to landscape-gardening, than to farm drainage. This, too, was found to be beyond the scope of our proposed treatise, and has been left to some abler hand.

So, too, the whole subject of reclaiming lands from the sea, and from rivers, by embankment, and the drainage of lakes and ponds, which at a future day must attract great attention in this country, has proved quite too extensive to be treated here. The day will soon come, when on our Atlantic coast, the ocean waves will be stayed, and all along our great rivers, the Spring floods, and the [19] Summer freshets, will be held within artificial barriers, and the enclosed lands be kept dry by engines propelled by steam, or some more efficient or economical agent.

The half million acres of fen-land in Lincolnshire, producing the heaviest wheat crops in England; and Harlaem Lake, in Holland, with its 40,000 acres of fertile land, far below the tides, and once covered with many feet of water, are examples of what science and well-directed labor may accomplish. But this department of drainage demands the skill of scientific engineers, and the employment of combined capital and effort, beyond the means of American farmers; and had we ability to treat it properly, would afford matter rather of pleasing speculation, than of practical utility to agricultural readers.

With a reckless expenditure of paper and ink, we had already prepared chapters upon several topics, which, though not essential to farm-drainage, were as near to our subject as the minister usually is limited in preaching, or the lawyer in argument; but conformity to the Procrustean bed, in whose sheets we had in advance stipulated to *sleep*, cost us the amputation of a few of our least important heads.

"Don't be too English," suggests a very wise and politic friend. We are fully aware of the prejudice which still exists in many minds in our country, against what is peculiarly English. Because, forsooth, our good Mother England, towards a century ago, like most fond mothers, thought her transatlantic daughter quite too young and inexperienced to set up an establishment and manage it for herself, and drove her into wasteful experiments of wholesale tea-making in Boston harbor, by way of illustrating her capacity of entertaining company from beyond seas; and because, near half a century ago, we had some sharp words, spoken not through the mouths of prophets and sages, but through the mouths of great guns, touching [20] the right of our venerated parent to examine the internal economy of our merchant-ships on the sea—because of reminiscences like these, we are to forswear all that is English! And so we may claim no kindred in literature with Shakspeare and Milton, in jurisprudence, with Bacon and Mansfield, in statesmanship, with Pitt and Fox!

Whence came the spirit of independence, the fearless love of liberty of which we boast, but from our English blood? Whence came our love of territorial extension, our national ambition, exhibited under the affectionate name of annexation? Does not this velvet paw with which we softly play with our neighbors' heads, conceal some long, crooked talons, which tell of the ancestral blood of the British Lion?

The legislature of a New England State, not many years ago, appointed a committee to revise its statutes. This committee had a pious horror of all dead languages, and a patriotic fear of paying too high a compliment to England, and so reported that all proceedings in courts of law should be in the American language! An inquiry by a waggish member, whether the committee intended to allow proceedings to be in any one of the three hundred Indian dialects, restored to the English language its appropriate name.

Though from some of our national traits, we might possibly be supposed to have sprung from the sowing of the dragon's teeth by Cadmus, yet the uniform record of all American families which goes back to the "three brothers who came over from England," contradicts this theory, and connects us by blood and lineage with that country.

Indeed, we can hardly consent to sell our birthright for so poor a mess of pottage as this petty jealousy offers. A teachable spirit in matters of which we are ignorant, is usually as profitable and respectable as abundant self-conceit, [21] and rendering to Cæsar the things that are Cæsar's, quite as honest as to pocket the coin as our own, notwithstanding the "image and superscription."

We make frequent reference to English writers and to English opinions upon our subject, because drainage is understood and practiced better in England than anywhere else in the world, and because by personal inspection of drainage-works there, and personal acquaintance and correspondence with some of the most successful drainers in that country, we feel some confidence of ability to apply English principles to American soil and climate.

To J. Bailey Denton, Engineer of the General Land Drainage Company, and one of the most distinguished practical and scientific drainers in England, we wish publicly to acknowledge our obligations for personal favors shown us in the preparation of our work.

We claim no great praise of originality in what is here offered to the public. Wherever we have found a person of whom we could learn anything, in this or other countries, we have endeavored to profit by his teachings, and whenever the language of another, in book or journal, has been found to express forcibly an idea which we deemed worthy of adoption, we have given full credit for both thought and words.

Our friends, Messrs. Shedd and Edson, of Boston, whose experience as draining engineers entitles them to a high rank among American authorities, have been in constant communication with us, throughout our labors. The chapter upon Evaporation, Rain fall, &c., which we deem of great value as a contribution to science in general, will be seen to be in part credited to them, as are also the tables showing the discharge of water through pipes of various capacity.

Drainage is a new subject in America, not well understood, [22] and we have no man, it is believed, peculiarly fitted to teach its theory and practice; yet the farmers everywhere are awake to its importance, and are eagerly seeking for information on the subject. Many are already engaged in the endeavor to drain their lands, conscious of their want of the requisite knowledge to effect their object in a profitable manner, while others are going resolutely forward, in violation of all correct principles, wasting their labor, unconscious even of their ignorance.

In New England, we have determined to dry the springy hill sides, and so lengthen our seasons for labor; we have found, too, in the valleys and swamps, the soil which has been washed from our mountains, and intend to avail ourselves of its fertility in the best manner practicable. On the prairies of the great West, large tracts are found just a little too wet for the best crops of corn and wheat, and the inquiry is anxiously made, how can we be rid of this surplus water.

There is no treatise, English or American, which meets the wants of our people. In England, it is true, land drainage is already reduced to a science; but their system has grown up by degrees, the first principles being now too familiar to be at all discussed, and the points now in controversy there, quite beyond the comprehension of beginners. America wants a treatise which shall be elementary, as well as thorough—that shall teach the alphabet, as well as the transcendentalism, of draining land—that shall tell the man who never saw a drain-tile what thorough drainage is, and shall also suggest to those who have studied the subject in English books only, the differences in climate and soil, in the prices of labor and of products, which must modify our operations.

With some practical experience on his own land, with careful observation in Europe and in America of the details of drainage operations, with a somewhat critical [23] examination of published books and papers on all topics connected with the general subject, the author has endeavored to turn the leisure hours of a laborious professional life to some account for the farmer. Although, as the lawyers say, the "presumptions" are, perhaps, strongly against the idea, yet a professional man *may* understand practical farming. The profession of the law has made some valuable contributions to agricultural literature. Sir Anthony Fitzherbert, author of the "Boke of Husbandrie," published in 1523, was Chief Justice of the Common Pleas, and, as he says, an *"experyenced farmer* of more than 40 years." The author of that charming little book, "Talpa," it is said, is also a lawyer, and there is such wisdom in the idea, so well expressed by Emerson as a fact, that we commend it by way of consolation to men of all the learned professions: "All of us keep the farm in reserve, as an asylum where to hide our poverty and our solitude, if we do not succeed in society."

Besides the prejudice against what is foreign, we meet everywhere the prejudice against what is new, though far less in this country than in England. "No longer ago than 1835," says the *Quarterly Review*, "Sir Robert Peel presented a Farmers' Club, at Tamworth, with two iron plows of the best construction. On his next visit, the old plows, with the wooden mould-boards, were again at work. 'Sir,' said a member of the club, 'we tried the iron, and we be all of one mind, that they make the weeds grow!'"

American farmers have no such ignorant prejudice as this. They err rather by having too much faith in themselves, than by having too little in the idea of progress, and will be more likely to "go ahead" in the wrong direction, than to remain quiet in their old position.

# CHAPTER II. [24]
# HISTORY OF THE ART OF DRAINING.

Draining as Old as the Deluge.—Roman Authors.—Walter Bligh in 1650.—No thorough drainage till Smith of Deanston.—No mention of tiles in the "Compleat Body of Husbandry," 1758.—Tiles found 100 years old.—Elkington's System.—Johnstone's Puns and Peripatetics.—Draining Springs.—Bletonism, or the Faculty of Perceiving Subterranean Water.—Deanston System.—Views of Mr. Parkes.—Keythorpe System.—Wharncliffe System.—Introduction of tiles into America.—John Johnston, and Mr. Delafield, of New York.

The art of removing superfluous water from land, must be as ancient as the art of cultivation; and from the time when Noah and his family anxiously watched the subsiding of the waters into their appropriate channels, to the present, men must have felt the ill effects of too much water, and adopted means more or less effective, to remove it.

The Roman writers upon agriculture, Cato, Columella, and Pliny, all mention draining, and some of them give minute directions for forming drains with stones, branches of trees, and straw. Palladius, in his *De Aquæ Ductibus*, mentions earthen-ware tubes, used however for aqueducts, rather for conveying water from place to place, than for draining lands for agriculture.

Nothing, however, like the systematic drainage of the present day, seems to have been conceived of in England, until about 1650, when Captain Walter Bligh published a work, which is interesting, as embodying and boldly [25] advocating the theory of deep-drainage as applied by him to water-meadows and swamps, and as applicable to the drainage of all other moist lands.

We give from the 7th volume of the Journal of the Royal Agricultural Society, in the language of that eminent advocate of deep-drainage, Josiah Parkes, an account of this rare book, and of the principles which it advocates, as a fitting introduction to the more modern and more perfect system of thorough drainage:

"The author of this work was a Captain Walter Bligh, signing himself, 'A Lover of Ingenuity.' It is quaintly entitled, 'The English

Improver Improved; or, the Survey of Husbandry Surveyed;' with several prefaces, but specially addressed to 'The Right Honorable the Lord General Cromwell, and the Right Honorable the Lord President, and the rest of the Honorable Society of the Council of State.' In his instructions for forming the flooding and draining trenches of water-meadows, the author says of the latter:—'And for thy drayning-trench, it must be made so deep, that it goe to the bottom of the cold spewing moyst water, that feeds the flagg and the rush; for the widenesse of it, use thine own liberty, but be sure to make it so wide as thou mayest goe to the bottom of it, which must be so low as any moysture lyeth, which moysture usually lyeth under the over and second swarth of the earth, in some gravel or sand, or else, where some greater stones are mixt with clay, under which thou must goe half one spade's graft deep at least. Yea, suppose this corruption that feeds and nourisheth the rush or flagg, should lie a yard or four-foot deepe; to the bottom of it thou must goe, if ever thou wilt drayn it to purpose, or make the utmost advantage of either floating or drayning, without which the water cannot have its kindly operation; for though the water fatten naturally, yet still this coldnesse and moysture lies gnawing within, and not being taken clean away, it eates out what the water fattens; and so the goodnesse of the water is, as it were, riddled, screened, and strained out into the land, leaving the richnesse and the leanesse sliding away from it.' In another place, he replies to the objectors of floating, that it will breed the rush, the flagg, and mare-blab; 'only make thy drayning-trenches deep enough, and not too far off thy floating course, and I'le warrant it they drayn away that under-moysture, fylth, and venom as aforesaid, that maintains them; and then believe me, or deny Scripture, which I hope thou doust not, as [26] Bildad said unto Job, "Can the rush grow without mire, or the flagg without water?" Job viii. 12. That interrogation plainly showes that the rush cannot grow, the water being taken from the root; for it is not the moystnesse upon the surface of the land, for then every shower should increase the rush, but it is that which lieth at the root, which, drayned away at the bottom, leaves it naked and barren of relief.'

"The author frequently returns to this charge, explaining over and over again the necessity of removing what we call bottom-water, and which he well designates as 'filth and venom.'

"In the course of my operations as a drainer, I have met with, or heard of, so many instances of swamp-drainage, executed precisely according to the plans of this author, and sometimes in a superior manner—the conduits being formed of walling stone, at a period long antecedent to the memory of the living—that I am disposed to consider the practice of deep drainage to have originated with Captain Bligh, and to have been preserved by imitators in various parts of the country; since a book, which passed through three editions in the time of the Commonwealth, must necessarily have had an extensive circulation, and enjoyed a high renown. Several complimentary autograph verses, written by some imitators and admirers of the ingenious Bligh, are bound up with the volume. I find also, not unfrequently, very ancient deep drains in arable fields, and some of them still in good condition; and in a case or two, I have met with several ancient drains six feet deep, placed parallel with each other, but at so great a distance asunder, as not to have commanded a perfect drainage of the intermediate space. The author from whom I have so largely quoted, is the earliest known to me, who has had the sagacity to distinguish between the transient effect of rain, and the constant action of stagnant bottom-water in maintaining land in a wet condition."

Dr. Shier, editor of "Davy's Agricultural Chemistry," says, "The history of drainage in Britain may be briefly told. Till the time of Smith, of Deanston, draining was generally regarded as the means of freeing the land from springs, oozes, and under-water, and it was applied only to lands palpably wet, and producing rushes and other aquatic plants."

He then proceeds to give the principles of Elkington, Smith, Parkes, and other modern writers, of which we shall speak more at large. [27]

The work published in England, not far from Captain Bligh's time, under the title "A Complete Body of Husbandry," undertakes to give directions for all sorts of farming processes. A Second Edition, in four octavo volumes, of which we have a copy, was published in 1758. It professes to treat of "Draining in General," and then of the draining of boggy land and of fens, but gives no intimation that any other lands require drainage.

Directions are given for filling drains with "rough stones," to be covered with refuse wood, and over that, some of the earth that was thrown out in digging. "By this means," says the writer, "a passage will be left free for all the water the springs yield, and there will be none of these great openings upon the surface."

He thus describes a method practiced in Oxfordshire of draining with bushes:

"Let the trenches be cut deeper than otherwise, suppose three foot deep, and two foot over. As soon as they are made, let the bottoms of them be covered with fresh-cut blackthorn bushes. Upon these, throw in a quantity of large refuse stones; over these let there be another covering of straw, and upon this, some of the earth, so as to make the surface level with the rest. These trenches will always keep open."

No mention whatever is made in this elaborate treatise of tiles of any kind, which affords very strong evidence that they were not in use for drainage at that time. In a note, however, to Stephen's "Draining and Irrigation," we find the following statement and opinion:

"In draining the park at Grimsthorpe, Lincolnshire, about three years ago, some drains, made with tiles, were found eight feet below the surface of the ground. The tiles were similar to what are now used, and in as good a state of preservation as when first laid, although they must have remained there above one hundred years."

## ELKINGTON'S SYSTEM OF DRAINAGE.

It appears, that, in 1795, the British Parliament, at the request of the Board of Agriculture, voted to Joseph [28] Elkington a reward of £1000, for his valuable discoveries in the drainage of land. Joseph Elkington was a Warwickshire farmer, and Mr. Gisborne says he was a man of considerable genius, but he had the misfortune to be illiterate. His discovery had created such a sensation in the agricultural world, that it was thought important to record its details; and, as Elkington's health was extremely precarious, the Board resolved to send Mr. John Johnstone to visit, in company with him, his principal works of drainage, and to transmit to posterity the benefits of his knowledge.

Accordingly, Mr. John Johnstone, having carefully studied Elkington's system, under its author, in the peripatetic method, undertook, like Plato, to record the sayings of his master in science, and produced a work, entitled, "An Account of the Most Approved Mode of Draining Land, According to the System Practised by Mr. Joseph Elkington." It was published at Edinburgh, in 1797. Mr. Gisborne says, that Elkington found in Johnstone "a very inefficient exponent of his opinions, and of the principles on which he conducted his works."

"Every one," says he, "who reads the work, which is popularly called 'Elkington on Draining,' should be aware, that it is not Joseph who thinks and speaks therein, but John, who tells his readers what, according to his ideas, Joseph would have thought and spoken."

Again—

"Johnstone, measured by general capacity, is a very shallow drainer! He delights in exceptional cases, of which he may have met with some, but of which, we suspect the great majority to be products of his own ingenuity, and to be put forward, with a view to display the ability with which he could encounter them."

Johnstone's report seems to have undergone several revisions, and to have been enlarged and reproduced in other forms than the original, for we find, that, in 1838, it was published in the United States, at Petersburg, Virginia, [29] as a supplement to the *Farmer's Register*, by Edmund Ruffin, Esq., editor, a reprint "from the third British Edition, revised and enlarged," under the following title:

"A Systematic Treatise on the Theory and Practice of Draining Land, &c., according to the most approved methods, and adapted to the various situations and soils of England and Scotland; also on sea, river, and lake embankments, formation of ponds and artificial pieces of water, with an appendix, containing hints and directions for the culture and improvement of bog, morass, moor, and other unproductive ground, after being drained; the whole illustrated by plans and sections applicable to the various situations and forms of construction. Inscribed to the Highland and Agricultural Society of Scotland, by John Johnstone, Land Surveyor."

Mr. Ruffin certainly deserves great credit for his enterprise in re-publishing in America, at so early a day, a work of which an English copy could not be purchased for less than six dollars, as well as for his zealous labors ever since in the cause of agriculture.

There is, in this work of Johnstone, a quaintness which he, proba-bly, did not learn from Elkington, and which illustrates the charac-ter of his mind as one not peculiarly adapted to a plain and practical history of another man's system and labors. For instance, in speak-ing of the arrangement of his subject into parts, he says, in a note, "The subject being closely connected with *cutting*, *section* is held as a better *division* than chapter!"

Again, he speaks of embanking, and says he has some experience on that head. Then he adds the following note, lest a possible pun should be lost: "An embankment is often termed a 'head,' as it makes head, or resistance, against the encroachment of high tide or river floods."

There is some danger that a mind which scents a whimsical anal-ogy of meaning like this, may entirely lose the main track of pursuit; but Johnstone's special mission [30] was to ascertain Elkington's method, and his account of it is, therefore, the best authority we have on the subject.

He gives the following statement of Elkington's discovery:

"In the year 1763, Elkington was left by his father in the posses-sion of a farm called Prince-Thorp, in the parish of Stretton-upon-Dunsmore, and county of Warwick. The soil of this farm was so poor, and, in many places, so extremely wet, that it was the cause of rotting several hundreds of his sheep, which first induced him, if possible, to drain it. This he begun to do, in 1764, in a field of wet clay soil, rendered almost a swamp, or *shaking* bog, by the springs which issued from an adjoining bank of gravel and sand, and over-flowed the surface of the ground below. To drain this field, which was of considerable extent, he cut a trench about four or five feet deep, a little below the upper side of the bog, where the wetness began to make its appearance; and, after proceeding with it in this direction and at this depth, he found it did not reach the *principal body of subjacent water* from which the evil arose. On perceiving this, he was at a loss how to proceed, when one of his servants came to the field with an *iron crow*, or bar, for the purpose of making holes for fixing sheep hurdles in an adjoining part of the farm, as represented on the plan. Having a suspicion that his drain was not deep enough, and desirous to know what strata lay under it, he took the iron bar, and having forced it down about four feet below the bottom of the trench, on pulling it out, to his astonishment, a great quantity of water burst up through the hole he had thus made, and ran along the drain. This led him to the knowledge, that wetness may be often produced by water confined farther below the surface of the ground than it was possible for the usual depth of drains to reach, and that an *auger* would be a useful instrument to apply in such cases. Thus, chance was the parent of this discovery, as she often is of other useful arts; and fortunate it is for society, when such accidents happen to those who have sense and judgment to avail themselves of hints thus fortuitously given. In this manner he soon accomplished the drainage of his whole farm, and rendered it so perfectly dry and sound, that none of his flock was ever after affected with disease.

"By the success of this experiment, Mr. Elkington's fame, as a drainer, was quickly and widely extended; and, after having successfully drained several farms in his neighborhood, he was, at last, very generally employed for that purpose in various parts of the kingdom, till about thirty years ago, when the country had the melancholy cause [31] to regret his loss. From his long practice and experience, he became so successful in the works he undertook, and so skillful in judging of the internal strata of the earth and the nature of springs, that, with remarkable precision, he could ascertain where to find water, and trace the course of springs that made no appearance on the surface of the ground. During his practice of more than thirty years, he drained in various parts of England, particularly in the midland counties, many thousand acres of land, which, from being originally of little or no value, soon became as useful as any in the kingdom, by producing the most valuable kinds of grain and feeding the best and healthiest species of stock.

"Many have erroneously entertained an idea that Elkington's skill lay solely in applying the auger for the *tapping of springs*, without attaching any merit to his method of conducting the drains. The accidental circumstance above stated gave him the first notion of using an auger, and directed his attention to the profession and practice of draining, in the course of which he made various useful discoveries, as will be afterwards explained. With regard to the use of the auger, though there is every reason to believe that he was led to employ that instrument from the circumstance already stated, and did not derive it from any other source of intelligence, yet there is no doubt that others might have hit upon the same idea without being indebted for it to him. It has happened, that, in attempts to discover mines by boring, springs have been tapped, and ground thereby drained, either by letting the water down, or by giving it vent to the surface; and that the auger has been likewise used in bringing up water in wells, to save the expense of deeper digging; but that it had been *used in draining land, before Mr. Elkington made that discovery, no one has ventured to assert.*"

Begging pardon of the shade of John Johnstone for the liberty, we will copy from Mr. Gisborne, as being more clearly expressed, a summary explanation of Elkington's system, as Mr. Gisborne has deduced it from Johnstone's report, with two simple and excellent plans:

"A slight modification of Johnstone's best and simplest plan, with a few sentences of explanation, will sufficiently elucidate Elkington's mystery, and will comprehend the case of all simple superficial springs. Perhaps in Agricultural Britain, no formation is more common than moderate elevations of pervious material, such as chalk, gravel, and imperfect stone or rock of various kinds, resting upon more horizontal [32] beds of clay, or other material less pervious than themselves, and at their inferior edge overlapped by it. For this overlap geological reasons are given, into which we cannot now enter. In order to make our explanation simple, we use the words, gravel and clay, as generic for pervious and impervious material.

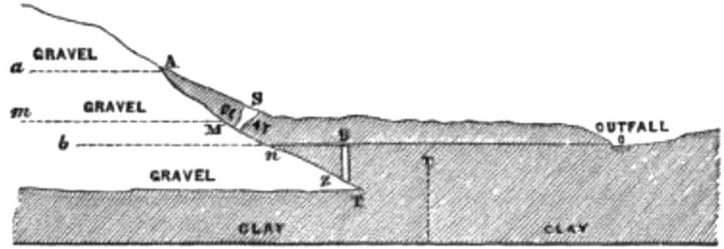

Fig. 1

"Our drawing is an attempt to combine plan and section, which will probably be sufficiently illustrative. From A to T is the overlap, which is, in fact, a dam holding up the water in the gravel. In this dam there is a weak place at S, through which water issues permanently (a superficial spring), and runs over the surface from S to O. This issue has a tendency to lower the water in the gravel to the line M $m$. But when continued rains overpower this issue, the water in the gravel rises to the line A $a$, and meeting with no impediment at the point A, it flows over the surface between A and S. In addition to these more decided outlets, the water is probably constantly squeezing, in a slow way, through the whole dam. Elkington undertakes to drain the surface from A to O. He cuts a drain from O to B, and then he puts down a bore-hole, an Artesian well, from B to Z. His hole enters the tail of the gravel; the water contained therein rises up it: and the tendency of this new outlet is to lower the water to the line B $b$. If so lowered that it can no longer overflow at A or at S, and the surface from A to O is drained, so far as the springs are concerned, though our section can only represent one spring, and one summit-overflow, it is manifest that, however long the horizontal line of junction between the gravel and clay may be, however numerous the weak places (springs) in the overlap, or dam, and the summit-overflows, they will all be stopped, provided they lie at a higher level than the line B $b$. If Elkington had driven his drain forward from B to $n$, he would, at least, equally have attained his object; but the bore-hole was [33] less expensive. He escapes the deepest and most costly portion of his drain. At $x$, he might have bored to the centre of the earth without ever realizing the water in this gravel. His whole success, therefore, depended upon his sagacity in hitting the point Z. Another simple and very common case, first successfully treated by Elkington, is illustrated by our second drawing.

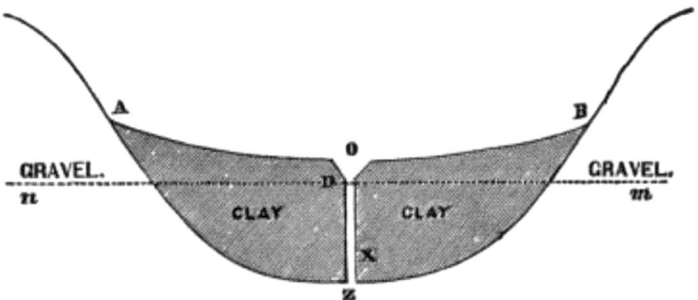

Fig. 2

"Between gravel hills lies a dish-shaped bed of clay, the gravel being continuous under the dish. Springs overflow at A and B, and wet the surface from A to O, and from B to O. O D is a drain four or five feet deep, and having an adequate outlet; D Z a bore-hole. The water in the gravel rises from Z to D, and is lowered to the level D *m* and D *n*. Of course it ceases to flow over at A and B. If Elkington's heart had failed him when he reached X, he would have done no good. All his success depends on his reaching Z, however deep it may lie. Elkington was a discoverer. We do not at all believe that his discoveries hinged on the accident that the shepherd walked across the field with a crow-bar in his hand. When he forced down that crow-bar, he had more in his head than was ever dreamed of in Johnstone's philosophy. Such accidents do not happen to ordinary men. Elkington's subsequent use of his discovery, in which no one has yet excelled him, warrants our supposition that the discovery was not accidental. He was not one of those prophets who are without honor in their own country: he created an immense sensation, and received a parliamentary grant of one thousand pounds. One writer compares his auger to Moses' rod, and Arthur Young speculates, whether though worthy to [34] be rewarded by millers on one side of the hill for increasing their stream, he was not liable to an action by those on the other for diminishing theirs."

Johnstone sums up this system as follows:

"Draining according to Elkington's principles depends chiefly upon three things:

"1. Upon discovering the main spring, or source of the evil.

"2. Upon taking the subterraneous bearings: and,

"3dly. By making use of the auger to reach and *tap* the springs, when the depth of the drain is not sufficient for that purpose.

"The first thing, therefore, to be observed is, by examining the adjoining high grounds, to discover what strata they are composed of; and then to ascertain, as nearly as possible, the inclination of these strata, and their connection with the ground to be drained, and thereby to judge at what place the level of the spring comes nearest to where the water can be cut off, and most readily discharged. The surest way of ascertaining the lay, or inclination, of the different strata, is, by examining the bed of the nearest streams, and the edges of the banks that are cut through by the water; and any pits, wells, or quarries that may be in the neighborhood. After the *main spring* has been thus discovered, the next thing is, to ascertain a line on the same level, to one or both sides of it, in which the drain may be conducted, which is one of the most important parts of the operation, and one on which the art of draining in a scientific manner essentially depends.

"Lastly, the use of the auger, which, in many cases, is the *sine qua non* of the business, is to reach and tap the spring when the depth of the drain does not reach it: where the level of the outlet will not admit of its being cut to a greater depth; and where the expense of such cutting would be great, and the execution of it difficult.

"According to these principles, this system of draining has been attended with extraordinary consequences, not only in laying the land dry in the vicinity of the drain, but also springs, wells, and wet ground, at a considerable distance, with which there was no apparent connection."

# DRAINAGE OF SPRINGS.

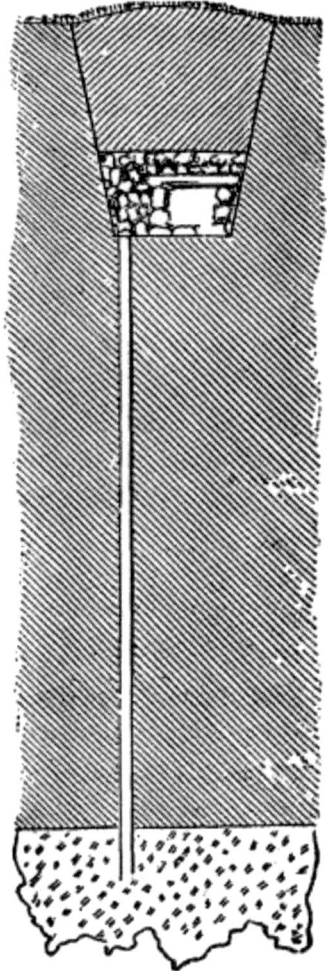

Fig. 3.

Wherever, from any cause, water bursts out from a hill's side, or from below, in a well defined spring, in any considerable quantity, the Elkington method of cutting a [35] deep drain directly into the seat of the evil, and so lowering the water that it may be carried away below the surface, is obviously the true and common-sense remedy. There may be cases where, in addition to the drain, it may be expedient to bore with an auger in the course of the drain. This, however, would be useful only where, from the peculiar formation, water is pent up upon a retentive subsoil in the manner already indicated. Elkington's method of draining by boring is illustrated in the following cut.

In studying the history of Elkington's discovery, and especially of his own application of it, it would seem that he must have possessed some peculiar faculty of ascertaining the subterranean currents of water, not possessed or even claimed by modern engineers.

Indeed, Mr. Denton, who may rightly claim as much skill as a draining engineer, perhaps, as any man in England, expressly says, "It does not appear that any person now will undertake to do what Elkington did sixty years back."

In the Patent Office Report for 1851, at page 14, may be found an article entitled, "Well-digging," in which it is gravely contended, and not without a fair show of evidence, that certain persons possess the power of indicating, by means of a sort of divining rod of hazel or willow, subterraneous currents or springs of water. [36] This power has been called Bletonism, which is defined by Webster to be, "the faculty of perceiving and indicating subterraneous springs and currents by sensation—so called from one Bleton, of France, who possessed this faculty."

Under the authority of Webster, and of Mr. Ewbank, the Commissioner of Patents, in whose report the article in question was published by the Government of the United States, it will not be considered, perhaps, as putting faith in "water-witchery," to suggest that, possibly, Elkington did really possess a faculty, not common to all mankind, of detecting running water or springs, even far below the surface. We have the high authority of Tam o' Shanter for the opinion, that witches cannot cross a stream of water; for, when pursued by the "hellish legion" from Kirk-Alloway, he put his "gude mare Meg" to do her "speedy utmost" for the bridge of Doon, knowing that,

"A running stream they darena cross."

If witches are thus affected by flowing water, there is no reason to doubt that others, of peculiar organization, may possess some sensitiveness at its presence.

It would not, probably, be useful to pursue more into detail the method of Mr. Elkington. The general principles upon which he wrought have been sufficiently explained. The miracles performed under his system seem to have ceased with his life, and, until we receive some new revelation as to the mode of finding the springs hidden in the earth, we must be content with the moderate results of a careful application of ordinary science, and not be discouraged in our attempts to leave the earth the better for our having lived on it, if we do not, like Elkington, succeed in draining, by a single ditch and a few auger holes, sixty statute acres of land.

## THE DEANSTON SYSTEM; OR, FREQUENT DRAINAGE. [37]

James Smith, Esq., of Deanston, Sterlingshire, in Scotland, next after Elkington, in point of time, is the prominent leader of drainage operations in Great Britain. His peculiar views came into general notice about 1832, and, in 1844, we find published a seventh edition of his "Remarks on Thorough Draining." Smith was a man of education, and seems to be, in fact, the first advocate of any system worthy the name of thorough drainage.

Instead of the few very deep drains, cut with reference to particular springs or sources of wetness, adopted by Elkington, Smith advocated and practiced a systematic operation over the whole field, at regular distances and shallow depths. Smith states, that in Scotland, much more injury arises from the retention of rain water, than from springs; while Elkington's attention seems to have been especially directed to springs, as the source of the evil.

The characteristic views of Smith, of Deanston, as stated by Mr. Denton, were:

"1st. *Frequent* drains at intervals of from ten to twenty-four feet.

"2nd. *Shallow* depth—not exceeding thirty inches—designed for the single purpose of freeing that depth of soil from stagnant and injurious water.

"3rd. '*Parallel drains at regular distances* carried throughout the whole field, without reference to the wet and dry appearance of portions of the field,' in order 'to provide frequent opportunities for the water, *rising from* below and falling on the surface, to pass freely and completely off.

"4th. *Direction of the minor drains* 'down the steep,' and that of the mains along the bottom of the chief hollow; tributary mains being provided for the lesser hollows.

"The reason assigned for the minor drains following the line of steepest descent, was, that 'the stratification generally lies in sheets at an angle to the surface.'

"5th. *As to material*—Stones preferred to tiles and pipes."

[38]

Mr. Smith somewhat modified his views during the last years of his life, especially as to the depth of drains, and, instead of shallow drains, recommended a depth of three feet, and even more in some cases; but continued, to the time of his death, which occurred about 1854, to oppose any increased intervals between the drains, and the extreme depth of four feet and more advocated by others. The peculiar points insisted on by Smith were, that drains should be near and parallel. His own words are:

"The drains should be parallel with each other and at regular distances, and should be carried throughout the whole field, without regard to the wet and dry appearance of portions of the field—the principle of this system being the providing of frequent opportunities for the water rising from below, or falling on the surface, to pass freely and completely off."

Mr. Smith called it the "frequent drain system," and Mr. Denton says, that, "for distinction sake, I have ventured to christen this ready-made practice, the *gridiron system*," a name, by the way, which will, probably, seem to most readers more distinctive than respectful. Whatever may be the improvements on the Deanston method of draining, the name of Mr. Smith deserves, and, indeed, has already obtained, a high place among the improvers of agriculture.

## VIEWS OF MR. PARKES.

About the year 1846, when the first Act of the British Parliament authorizing "the advance of public money to promote the improvement of land by works of drainage" was passed, a careful investigation of the whole subject was made by a Committee of the House of Lords, and it was found that the best recorded opinions, if we except the peculiar views of Elkington, were represented by, if not merged into, those of Smith, of Deanston, which have already been stated, or those of Josiah Parkes. Mr. Parkes is the author of "Essays on the Philosophy and [39] Art of Land Drainage," and of many valuable papers on the same subject, published in the journal of the Royal Agricultural Society, of which he was consulting engineer. He is spoken of by Mr. Denton as "one whose philosophical publications on the same subject gave a scientific bearing to it, quite irreconcilable with the more mechanical rules laid down by Mr. Smith."

The characteristic views of Mr. Parkes, as set forth at that time, as compared with those of Mr. Smith, are—

"1st. *Less frequent drains*, at intervals varying from twenty-one to fifty feet, *with preference for wide intervals*.

"2nd. *Deeper drains at a minimum depth of four feet*, designed with the two-fold object of not only freeing the active soil from stagnant and injurious water, but of converting the water falling on the surface into an agent for fertilizing; no drainage being deemed efficient that did not both remove the water failing on the surface, and 'keep down the subterranean water at a depth exceeding the power of capillary attraction to elevate it to near the surface.'

"3rd. *Parallel arrangement of drains*, as advocated by Smith, of Deanston.

"4th. *The advantage of increased depth*, as compensating for increased width between the drains.

"5th. *Pipes of an inch bore, the 'best known conduit'* for the parallel drains. (See Evidence before Lords' Committee on Entailed Estates, 1845, Q. 67.)

"6th. *The cost of draining uniform clays should not exceed £3 per acre.*"

The most material differences between the views of these two leaders of what have been deemed rival systems of drainage, will be seen to be the following. Smith advocates drains of two to three feet in depth, at from ten to twenty-four feet distances; while Parkes contends for a depth of not less than four feet, with a width between of from twenty-one to fifty feet, the depth in some measure compensating for the increased distance.

Mr. Parkes advocated the use of pipes of *one* inch bore, [40] which Mr. Smith contemptuously denominated "pencil-cases," and which subsequent experience has shown to be quite too small for prudent use.

The estimate of Mr. Parkes, based, in part, upon his wide distances and small pipes, that drainage might be effected generally in England at a cost of about fifteen dollars per acre, was soon found to be far below the average expense, which is now estimated at nearly double that sum.

The Enclosure Commissioners, after the most careful inquiry, adopted fully the views of Mr. Parkes as to the *depth* of drains. Mr. Parkes himself, saw occasion to modify his ideas, as to the cost of drainage, upon further investigation of the subject, and fixed his estimates as ranging from $15 to $30 per acre, according to soil and other local circumstances.

It has been well said by a recent English writer, of Mr. Parkes:

"That gentleman's services in the cause of drainage, have been inestimable, and his high reputation will not be affected by any remarks which experience may suggest with reference to details, so long as the philosophical principles he first advanced in support of deep drainage are acknowledged by thinking men. Mr. Parkes' practice in 1854, will be found to differ very considerably from his anticipations of 1845, but the influence of his earlier writings and sayings continues to this day."

## THE KEYTHORPE SYSTEM.

Lord Berners having adopted a method of drainage on his estate at *Keythorpe*, differing somewhat from any of the regular and more uniform modes which have been considered, a sharp controversy as to its merits has arisen, and still continues in England, which, like most controversies, may be of more advantage to others than to the parties immediately concerned.

The theory of the Keythorpe system seems to have been invented by Mr. Joshua Trimmer, a distinguished geologist [41] of England, who, about 1854, produced a paper, which was published in the journal of the Royal Agricultural Society, on the "Keythorpe System." He states that his own theory was based entirely on his knowledge of the geological structure of the earth, which will be presently given in his own language, and that he afterwards ascertained that Lord Berners, who had no special theory to vindicate, had, by the "tentative process," or in plain English, by trying experiments, hit upon substantially the same system, and found it to work admirably.

Most people in the United States have no idea of what it is to be patronized by a lord. In England, it is thought by many to be the thing needful to the chance, even, of success of any new theory, and accordingly, Mr. Trimmer, without hesitation, availed himself of the privilege of being patronized by Lord Berners; and the latter, before he was aware of how much the agricultural world was indebted to him for his valuable discoveries, suddenly found himself at the head of the "Keythorpe System of Drainage."

His lordship was probably as much surprised to ascertain that he had been working out a new system, as some man of whom we have heard, was, to learn that he had been speaking *prose* all his life! At the call of the public, however, his lordship at once gave to the world the facts in his possession, making no claim to any great discovery, and leaving Mr. Trimmer to defend the new system as best he might. The latter, in one of his pamphlets published in defence of the Keythorpe system, states its claims as follows:

"The peculiarities of the Keythorpe system of draining consist in this—that the parallel drains are not equidistant, and that they cross the line of the greatest descent. The usual depth is three and a half feet, but some are as deep as five and six feet. The depth and width of interval are determined by digging trial-holes, in order to ascertain not only the depth at which the bottom water is reached, but the height [42] to which the water rises in the holes, and the distance at which a drain will lay the hole dry. In sinking these holes, clay-banks are found with hollows or furrows between them, which are filled with a more porous soil, as represented in the annexed sectional diagram.

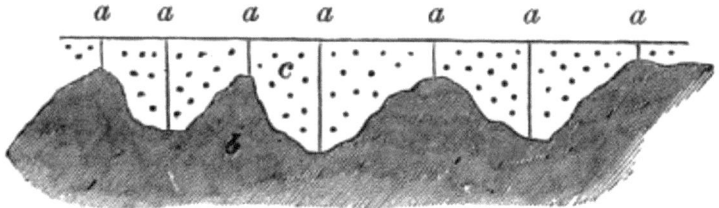

Fig. 4.

- *a a* Trial-holes.
- *b* Clay-banks of lias or of boulder-clay.
- *c* A more porous warp-drift filling furrows between the clay-banks.

"The next object is to connect these furrows by drains laid across them. The result is, that as the furrows and ridges here run along the fall of the ground, which I have observed to be the case generally elsewhere, the sub-mains follow the fall, and the parallel drains cross it obliquely.

"The intervals between the parallel drains are irregular, varying, in the same field, from 14 to 21, 31, and 59 feet. The distances are determined by opening the diagonal drains at the greatest distance from the trial-holes at which experience has taught the practicability of its draining the hole. If it does not succeed in accomplishing the object, another drain is opened in the interval. It has been found, in many cases, that a drain crossing the clay-banks and furrows takes the water from holes lying lower down the hill; that is to say, it intercepts the water flowing to them through these subterranean channels. The parallel drains, however, are not invariably laid across the fall. The exceptions are on ground where the fall is very slight, in which case they are laid along the line of greatest descent. On such grounds there are few or no clay-banks and furrows."

It would seem highly probable that the mode of drainage adopted at Keythorpe, is indebted for its success at that place, to a geological formation not often met with. At a public discussion in England, Mr. T. Scott, a gentleman of large experience in draining, stated that "he never, in his practice, had met with such a geological [43] formation as was said to exist at Keythorpe, except in such large areas as to admit of their being drained in the usual *gridiron* or parallel fashion."

It is claimed for this system by its advocates, that it is far cheaper than any other, because drains are only laid in the places where, by careful examination beforehand, by opening pits, they are found to be necessary; and that is a great saving of expense, when compared with the system of laying the drains at equal distances and depths over the field.

Against what is urged as the Keythorpe system, several allegations are brought.

In the first place, that it is in fact *no system*. Mr. Denton, having carefully examined the Keythorpe estate, and the published statements of its owner, asserts, that the drains there laid have *no uniformity of depth* — part of the tiles being laid but eighteen inches deep, and others four feet and more, in the same field.

Secondly, that there is *no uniformity as to direction* — part of the drains being laid across the fall, and part with the fall, in the same fields — with no obvious reason for the difference of direction.

Thirdly, that there is *no uniformity as to materials* — a part of the drains being wood, and a part tiles, in the same field.

Finally, it is contended that there is no saving of expense in the Keythorpe draining, over the ordinary mode, when all points are considered, because the pretended saving is made by the use of wood, where true economy would require tiles, and shallow drains are used where deeper ones would in the end be cheaper.

In speaking of this controversy, it is due to Lord Berners to say, that he expressly disclaims any invention or novelty in his operations at Keythorpe.

On the whole, although a work at the present day [44] which should pass over, without consideration, the claims of the Keythorpe system, would be quite incomplete in its history of the subject, yet the facts elicited with regard to it are perhaps chiefly valuable, as tending to show the danger of basing a general principle upon an isolated case.

The discussion of the claims of that system—if such it may be called—may be valuable in America, where novelty is sure to attract, by showing that one more form of error has already been tried and "found wanting;" and so save us the trouble of proving its inutility by experiment.

## THE WHARNCLIFFE SYSTEM.

Lord Wharncliffe, with a view to effect adequate drainage at less expense than is usual in thorough drainage, has adopted upon his estate a sort of compromise system, which he has brought to the notice of the public in the Journal of the Royal Agricultural Society.

Upon Fontenelle's idea, that "mankind only settle into the right course after passing through and exhausting all the varieties of error," it is well to advise our readers of this particular form of error also—to show that it has already been tried—so that no patent of invention can be claimed upon it by those perverse persons who are not satisfied without constant change, and who seem to imagine that the ten commandments might be improved by a new edition.

Lord Wharncliffe states his principles as follows, and calls his method the combined system of deep and shallow drainage:

"In order to secure the full effect of thorough drainage in clays, it is necessary that there should be not only well-laid conduits for the water which reaches them, but also subsidiary passages opened through the substance of the close subsoil, by means of atmospheric heat, and the contraction which ensues from it. The cracks and fissures which result [45] from this action, are reckoned upon as a certain and essential part of the process.

"To give efficiency, therefore, to a system of deep drains beneath a stiff clay, these natural channels are required. To produce them, there must be a continued action of heat and evaporation. If we draw off effectually and constantly the bottom water from beneath the clay and from its substance, as far as it admits of percolation, and by some other means provide a vent for the upper water, which needs no more than this facility to run freely, there seems good reason to suppose that the object may be completely attained, and that we shall remove the moisture from both portions as effectually as its quantity and the substance will permit. Acting upon this view, then, after due consideration, I determined to combine with the fundamental four-feet drains a system of auxiliary ones of much less depth, which should do their work above, and contribute their share to the wholesome discharge, while the under-current from their more subterranean neighbors should be steadily performing their more difficult duty.

"I accomplished this, by placing my four-feet drains at a distance of from eighteen to twenty yards apart, and then leading others into them, sunk only to about two feet beneath the surface (which appeared, upon consideration, to be sufficiently below any conceivable depth of cultivation), and laying these at a distance from each other of eight yards. These latter are laid at an acute angle with the main-drains, and at their mouths are either gradually sloped downwards to the lower level, or have a few loose stones placed in the same intervals between the two, sufficient to ensure the perpendicular descent of the upper stream through that space, which can never exceed, or, indeed, strictly equal, the additional two feet."

There are two reasons why this mode of drainage cannot be adopted in the northern part of the United States.

First: The two-foot drains would be liable to be frozen up solid, every winter.

Secondly: The subsoil plow, now coming into use among our best cultivators, runs to so great a depth as to be likely to entirely destroy two-foot drains at the first operation, even if it were not intended to run the sub-soiler to a greater general depth than eighteen inches. Any one who has had experience in holding a subsoil-plow, [46] must know that it is an implement somewhat unmanageable, and liable to plunge deep into soft spots like the covering over drains; so that no skill or care could render its use safe over two-foot drains.

The history of drainage in America, is soon given. It begins here, as it must begin everywhere, when practiced as a general system, with the introduction of tiles.

In 1835, Mr. John Johnston, of Seneca County, New York, a Scotchman by birth, imported from Scotland patterns of drain-tiles, and caused them to be made by hand-labor, and set the example of their use on his own farm. The effects of Mr. Johnston's operations were so striking, that in 1848, John Delafield, Esq., for a long time President of the Seneca County Agricultural Society, imported from England one of Scragg's Patent Tile machines. From that time, tile-draining in that county, and in the neighboring counties, has been diligently and profitably pursued. Several interesting statements of successful experiments by Mr. Johnston, Mr. Delafield, Mr. Theron G. Yeomans of Wayne County, and others, have been published, from time to time, in the "New York Transactions." Indeed, most of our information of experimental draining in this country, has come from that quarter.

Mr. Johnston, for more than twenty years, has made himself useful to the country, and at the same time gained a wide reputation for himself, by occasional publications on the subject of drainage.

In addition to this, his practical knowledge of agriculture, and especially of the subject of drainage, has gained for him a competence for his declining years. In this we rejoice; and trust that in these, his latter years, he may be made ever to feel, that even they among us of the friends of agriculture who have not known him personally, are not unmindful of their obligations to him as the leader of a most beneficent enterprise. [47]

Tile-works have since been established at various places in New York, at several places in Massachusetts, Ohio, Michigan, and many other States. The first drain-tiles used in New-Hampshire, were brought from Albany, in 1854, by Mr. William Conner, and used on his farm in Exeter, that year; and the following year, the writer brought some from Albany, and laid them on his farm, in the same town.

In 1857, tile-works were put in operation at Exeter; and some 40,000 tiles were made that year.

The horse-shoe tiles, we understand, have been generally used in New York. At Albany, and in Massachusetts, the sole-tile has been of late years preferred. We cannot learn that cylindrical pipes have ever been manufactured in this country until the Summer of 1858 when the engineers of the New York Central Park procured them to be made, and laid them, with collars, in their drainage-works there. This is believed to be the first practical introduction into this country of round pipes and collars, which are regarded in England as the most perfect means of drainage.

Experiments all over the country, in reclaiming bog-meadows, and in draining wet lands with drains of stone and wood, have been attempted, with various success.

Those attempts we regard as merely efforts in the right direction, and rather as evidence of a general conviction of the want, by the American farmer, of a cheap and efficient mode of drainage, than as an introduction of a system of thorough drainage; for—as we think will appear in the course of this work—no system of drainage can be made sufficiently cheap and efficient for general adoption, with other materials than drain-tiles.

# CHAPTER III [48]
# RAIN, EVAPORATION, AND FILTRATION.

Fertilizing Substances in Rain Water.—Amount of Rain Fall in United States—in England.—Tables of Rain Fall.—Number of Rainy Days, and Quantity of Rain each Month.—Snow, how Computed as Water.—Proportion of Rain Evaporated.—What Quantity of Water Dry Soil will Hold.—Dew Point.—How Evaporation Cools Bodies.—Artificial Heat Underground.—Tables of Filtration and Evaporation.

Although we usually regard drainage as a means of rendering land sufficiently dry for cultivation, that is by no means a comprehensive view of the objects of the operation.

Rain is the principal source of moisture, and a surplus of moisture is the evil against which we contend in draining. But rain is also a principal source of fertility, not only because it affords the necessary moisture to dissolve the elements of fertility already in the soil, but also because it contains in itself, or brings with it from the atmosphere, valuable fertilizing substances. In a learned article by Mr. Caird, in the Cyclopedia of Agriculture, on the Rotation of Crops, he says:

"The surprising effects of a fallow, even when unaided by any manure, has received some explanation by the recent discovery of Mr. Barral, that rain-water contains within itself, and conveys into the soil, fertilizing substances of the utmost importance, equivalent, in a fall of rain of 24 inches per annum, to the quantity of ammonia contained in 2 cwt. of Peruvian guano, with 150 lbs. of nitrogeneous matter besides, all suited to the nutrition of our crops."

[49]

About 42 inches of rain may be taken as a fair general average of the rain-fall in the United States. If this supplies as much ammonia to the soil as 3 cwt. of Peruvian guano to the acre, which is considered a liberal manuring, and which is valuable principally for its ammonia, we at once see the importance of retaining the rain-water long enough upon our fields, at least, to rob it of its treasures. But rain-water has a farther value than has yet been suggested:

"Rain-water always contains in solution, air, carbonic acid, and ammonia. The two first ingredients are among the most powerful disintegrators of a soil. The oxygen of the air, and the carbonic acid being both in a highly condensed form, by being dissolved, possess very powerful affinities for the ingredients of the soil. The oxygen attacks and oxydizes the iron; the carbonic acid seizing the lime and potash and other alkaline ingredients of the soil, produces a further disintegration, and renders available the locked-up ingredients of this magazine of nutriment. Before these can be used by plants, they must be rendered soluble; and this is only affected by the free and renewed access of rain and air. The ready passage of both of these, therefore, enables the soil to yield up its concealed nutriment."

We see, then, that the rains of heaven bring us not only water, but food for our plants, and that, while we would remove by proper drainage the surplus moisture, we should take care to first conduct it through the soil far enough to fulfill its mission of fertility. We cannot suppose that all rain-water brings to our fields precisely the same proportion of the elements of fertility, because the foreign properties with which it is charged, must continually vary with the condition of the atmosphere through which it falls, whether it be the thick and murky cloud which overhangs the coal-burning city, or the transparent ether of the mountain tops. We may see, too, by the tables, that the quantity of rain that falls, varies much, not only with the varying seasons of the year, and with the different seasons of different years, but with the distance [50] from the equator, the diversity of mountain and river, and lake and wood, and especially with locality as to the ocean. Yet the average results of nature's operations through a series of years, are startlingly constant and uniform, and we may deduce from tables of rain-falls, as from bills of mortality and tables of longevity, conclusions almost as reliable as from mathematical premises.

The quantity of rain is generally increased by the locality of mountain ranges. "Thus, at the Edinburgh Water Company's works, on the Pentland Hills, there fell in 1849, nearly twice as much rain as at Edinburgh, although the distance between the two places is only seven miles."

Although a much greater quantity of rain falls in mountainous districts (within certain limits of elevation) than in the plains, yet a greater quantity of rain falls at the surface of the ground than at an elevation of a few hundred feet. Thus, from experiments which were carefully made at York, it was ascertained that there fell eight and a half inches more rain at the surface of the ground, in the course of twelve months, than at the top of the Minster, which is 212 feet high. Similar results have been obtained in many other places.

Some observations upon this point may also be found in the Report of the Smithsonian Institution for 1855, at p. 210, given by Professor C. W. Morris, of New York.

Again, the evaporation from the surface of water being much greater than from the land, clouds that are wafted by the winds from the sea to the land, condense their vapor upon the colder hills and mountain sides, and yield rain, so that high lands near the sea or other large bodies of water, from which the winds generally blow, have a greater proportion of rainy days and a greater fall of rain than lands more remote from water. The annual rain-fall in the lake districts in Cumberland County, in England, sometimes amounts to more than 150 inches. [51]

With a desire to contribute as much as possible to the stock of accurate knowledge on this subject, we availed ourselves of the kindly offered services of our friends, Shedd and Edson, in preparing a carefully considered article upon a part of our general subject, which has much engaged their attention. Neither the article itself, nor the observations of Dr. Hobbs, which form a part of its basis, has ever before been published, and we believe our pages cannot be better occupied than by giving them in the language of our friends:

"All vegetables, in the various stages of growth, require warmth, air, and moisture, to support life and health.

Below the surface of the ground there is a body of stagnant water, sometimes at a great depth, but in retentive soils usually within a foot or two of the surface. This stagnant water not only excludes the air, but renders the soil much colder, and, being in itself of no benefit, without warmth and air, its removal to a greater depth is very desirable.

A knowledge of the depth to which this water-table should be removed, and of the means of removing it, constitutes the science of draining, and in its discussion, a knowledge of the rain-fall, humidity of the atmosphere, and amount of evaporation, is very important.

The amount of rain-fall, as shown by the hyetal, or rain-chart, of North America, by Lorin Blodget, is thirty inches vertical depth in the basin of the great lakes; thirty-two inches on Lake Erie and Lake Champlain; thirty-six inches in the valley of the Hudson, on the head waters of the Ohio, through the middle portions of Pennsylvania and Virginia, and western portion of North Carolina; forty inches in the extreme eastern and the northern portion of Maine, northern portions of New Hampshire and Vermont, south-eastern counties of Massachusetts, [52] Central New York, north-east portion of Pennsylvania, south-east portion of New Jersey and Delaware; also, on a narrow belt running down from the western portion of Maryland, through Virginia and North Carolina, to the north-western portion of South Carolina; thence, up through the western portion of Virginia, north-east portion of Ohio, Northern Indiana and Illinois, to Prairie du Chien; forty-two inches on the east coast of Maine, Eastern Massachusetts, Rhode Island, and Connecticut, and middle portion of Maryland; thence, on a narrow belt to South Carolina; thence, up through Eastern Tennessee, through Central Ohio, Indiana, and Illinois, to Iowa; thence, down through Western Missouri and Texas to the Gulf of Mexico; forty-five inches from Concord, New Hampshire, through Worcester, Mass., Western Connecticut, and the City of New York, to the Susquehanna River, just north of Maryland; also, at Richmond, Va., Raleigh, N. C., Augusta, Geo., Knoxville, Tenn., Indianopolis, Ind., Springfield, Ill., St. Louis, Mo.; thence, through Western Arkansas, across Red River to the Gulf of Mexico. From the belt just described, the rain-fall increases inland and southward, until at Mobile, Ala., the rain-fall is sixty-three inches. The same amount also falls in the extreme southern portion of Florida.

In England, the average rain-fall in the eastern portion is represented at twenty inches; in the middle portion, twenty-two inches; in the southern and western, thirty inches; in the extreme south-western, forty-five inches; and in Wales, fifty inches. In the eastern portion of Ireland, it is twenty-five inches; and in the western, forty inches.

Observations at London for forty years, by Dalton, gave average fall of 20.69 inches. Observations at New Bedford, Mass., for forty-three years, by S. Rodman, gave average fall of 41.03 inches — about double the amount in [53] London. The mean quantity for each month, at both places, is as follows:

|  | *New Bedford.* | *London.* |
|---|---|---|
| January | 3.36 | 1.46 |
| February | 3.32 | 1.25 |
| March | 3.44 | 1.17 |
| April | 3.60 | 1.28 |
| May | 3.63 | 1.64 |
| June | 2.71 | 1.74 |
| July | 2.86 | 2.45 |
| August | 3.61 | 1.81 |
| September | 3.33 | 1.84 |
| October | 3.46 | 2.09 |
| November | 3.97 | 2.22 |
| December | 3.74 | 1.74 |
| Spring | 10.67 | 4.09 |
| Summer | 9.18 | 6.00 |
| Autumn | 10.76 | 6.15 |
| Winter | 10.42 | 4.45 |

| Year | | | | | 41.03 | | 20.69 |
|---|---|---|---|---|---|---|---|

Another very striking difference between the two countries is shown by a comparison of the quantity of water falling in single days. The following table, given in the Radcliffe Observatory Reports, Oxford, England, 15th volume, shows the proportion of very light rains there. The observation was in the year 1854. Rain fell on 156 days:

| 73 | days gave | less than | | | .05 | inch. |
|---|---|---|---|---|---|---|
| 30 | " | between | that | and | .10 | " |
| 27 | " | between | .10 | " | .20 | " |
| 9 | " | " | .20 | " | .30 | " |
| 9 | " | " | .30 | " | .40 | " |
| 4 | " | " | .40 | " | .50 | " |
| 1 | gave | | | | .60 | " |
| 2 | " | | | | .80 | " |
| 1 | " | | | | 1.00 | " |

[54] Nearly half the number gave less fall than five-hundredths of an inch, and more than four-fifths the number gave less than one-fifth of an inch, and none gave over an inch.

There is more rain in the United States, by a large measure, than there; but the amount falls in less time, and the average of saturation is certainly much less here. From manuscript records, furnished us by Dr. Hobbs, of Waltham, Mass., we find, that the quantity falling in the year 1854, was equal to the average quantity for thirty years, and that rain fell on fifty-four days, in the proportion as follows:

Number of rainy days, 54; total rain-fall, 41.29.

| 0 | days gave | less than | | | .05 | inch. |
|---|---|---|---|---|---|---|

| | | | | | | |
|---|---|---|---|---|---|---|
| 2 | " | between | that | and | .10 | " |
| 8 | " | between | .10 | " | .20 | " |
| 7 | " | " | .20 | " | .30 | " |
| 5 | " | " | .30 | " | .40 | " |
| 4 | " | " | .40 | " | .50 | " |
| 2 | " | " | .50 | " | .60 | " |
| 4 | " | " | .60 | " | .70 | " |
| 4 | " | " | .70 | " | .80 | " |
| 3 | " | " | .80 | " | .90 | " |
| 0 | " | " | .90 | " | 1.00 | " |
| 0 | " | " | 1.00 | " | 1.10 | " |
| 2 | " | " | 1.10 | " | 1.20 | " |
| 1 | " | " | 1.20 | " | 1.30 | " |
| 1 | " | " | 1.30 | " | 1.40 | " |
| 3 | " | " | 1.40 | " | 1.50 | " |
| 2 | " | " | 1.50 | " | 1.60 | " |
| 1 | " | " | 1.60 | " | 1.70 | " |
| 2 | " | " | 1.80 | " | 1.90 | " |
| 1 | " | " | 2.30 | " | 2.40 | " |
| 1 | " | " | 2.50 | " | 2.60 | " |
| 1 | " | " | 3.20 | " | 3.30 | " |

No rain-fall gave less than five-hundredths of an inch; and more than one-fourth the number of days gave more [55] than one inch. In 1850, four years earlier, the rain-fall for the year, in Waltham, was 62.13 inches, the greatest recorded by observations kept since 1824. It fell as shown in the table:

Number of rainy days, 58; total rain-fall, 62.13.

| | | | | | |
|---|---|---|---|---|---|
| 3 | days gave between | .05 | and | .10 | inches. |
| 4 | " | .10 | " | .20 | " |
| 6 | " | .20 | " | .30 | " |
| 3 | " | .30 | " | .40 | " |
| 5 | " | .40 | " | .50 | " |
| 3 | " | .50 | " | .60 | " |
| 3 | " | .60 | " | .70 | " |
| 3 | " | .70 | " | .80 | " |
| 2 | " | .80 | " | .90 | " |
| 1 | " | .90 | " | 1.00 | " |
| 3 | " | 1.00 | " | 1.10 | " |
| 7 | " | 1.20 | " | 1.30 | " |
| 2 | " | 1.80 | " | 1.90 | " |
| 2 | " | 1.90 | " | 2.00 | " |
| 3 | " | 2.00 | " | 2.10 | " |
| 2 | " | 2.10 | " | 2.20 | " |
| 1 | " | 2.30 | " | 2.40 | " |
| 1 | " | 2.50 | " | 2.60 | " |

| 1 | " | 2.60 | " | 2.70 | " |
|---|---|------|---|------|---|
| 1 | " | 2.80 | " | 2.90 | " |
| 1 | " | 3.60 | " | 3.70 | " |
| 1 | " | 4.50 | " | 4.60 | " |

Sept. 7th and 8th, in 24 hours, 6.88 inches of rain fell, the greatest quantity recorded in one day.

In 1846—still earlier by four years—the rain-fall in Waltham was 26.90 inches—the least recorded by the same observations. It fell, as shown in the table: Number of rainy days, 49; total rain-fall, 26.90.

| 3 | days gave between | .05 | and | .10 | inches. |
|---|---|---|---|---|---|
| 7 | " | .10 | " | .20 | " |
| 10 | " | .20 | " | .30 | " |
| 6 | " | .30 | " | .40 | " |
| [56] 4 | " | .40 | " | .50 | " |
| 3 | " | .50 | " | .60 | " |
| 2 | " | .70 | " | .80 | " |
| 3 | " | .80 | " | .90 | " |
| 1 | " | .90 | " | 1.00 | " |
| 3 | " | 1.00 | " | 1.10 | " |
| 2 | " | 1.10 | " | 1.20 | " |
| 1 | " | 1.20 | " | 1.30 | " |
| 2 | " | 1.40 | " | 1.50 | " |
| 1 | " | 1.50 | " | 1.60 | " |

| 1 | " | 2.40 | " | 2.50 | " |
|---|---|---|---|---|---|

The rain-fall in 1852 was very near the average for thirty years; and the quantity falling in single storms, on sixty-three different occasions, as registered by Dr. Hobbs, was as follows: Number of storms, 63; total rain-fall, 42.24.

| 7 | storms gave | less than | | | .10 | inches. |
|---|---|---|---|---|---|---|
| 11 | " | between | .10 | and | .20 | " |
| 9 | " | " | .20 | " | .30 | " |
| 5 | " | " | .30 | " | .40 | " |
| 6 | " | " | .40 | " | .50 | " |
| 5 | " | " | .50 | " | .60 | " |
| 1 | " | " | .60 | " | .70 | " |
| 1 | " | " | .70 | " | .80 | " |
| 3 | " | " | .80 | " | .90 | " |
| 1 | " | " | .90 | " | 1.00 | " |
| 5 | " | " | 1.00 | " | 1.10 | " |
| 1 | " | " | 1.10 | " | 1.20 | " |
| 1 | " | " | 1.20 | " | 1.30 | " |
| 1 | " | " | 1.40 | " | 1.50 | " |
| 3 | " | " | 1.60 | " | 1.70 | " |
| 1 | " | in 5 days | | | 3.16 | " |
| 1 | " | " 4 " | | | 4.38 | " |
| 1 | " | " 6 " | | | 5.35 | " |

These tables are sufficient to show that provision must be made to carry off much greater quantities of water from lands in this country than in England. We add a table of the greatest fall of rain in any one day, for each month, and for the year, from April, 1824, to 1st January, [57] 1859. It also was abstracted from the manuscript of observations by Dr. Hobbs, and will be, we think, quite useful:

| Years | January | February | March | April | May | June | July | August | September | O |
|---|---|---|---|---|---|---|---|---|---|---|
| 1824 | | | | 0.76 | 0.67 | 0.53 | 0.44 | 1.90 | 2.54 | |
| 1825 | 2.16 | | 2.61 | 0.27 | 1.23 | 1.37 | 0.91 | 2.51 | 0.89 | |
| 1826 | 1.80 | 0.56 | 1.67 | 0.89 | 0.39 | 1.78 | 0.87 | 1.80 | 1.57 | |
| 1827 | | | 3.81 | 1.55 | 2.42 | 0.66 | 1.36 | 3.16 | 4.93 | |
| 1828 | 0.60 | 1.48 | 1.82 | 2.06 | 2.01 | 1.44 | 1.52 | 0.14 | 1.82 | |
| 1829 | 3.86 | 1.98 | 4.12 | 2.35 | 1.15 | 0.97 | 1.92 | 0.97 | 1.39 | |
| 1830 | 1.31 | | 1.17 | 2.68 | 2.28 | 0.78 | 1.84 | 2.45 | 2.40 | |
| 1831 | 0.64 | 1.48 | 2.32 | 2.12 | 1.79 | 1.87 | 2.27 | 1.00 | 1.00 | |
| 1832 | 2.68 | 1.59 | 2.00 | 4.48 | 2.52 | 1.24 | | 2.13 | 0.80 | |
| 1833 | 0.83 | | | 2.57 | 0.98 | 2.03 | 1.42 | 0.64 | 2.75 | |
| 1834 | | 0.64 | 1.31 | 0.94 | 2.35 | 1.87 | 2.12 | 0.73 | 1.25 | |
| 1835 | 1.44 | 0.88 | 2.48 | 2.48 | 1.18 | 1.52 | 4.72 | 1.32 | 1.57 | |
| 1836 | 2.72 | 3.04 | 2.26 | 1.86 | 1.29 | 2.24 | 1.04 | 0.72 | 0.36 | |
| 1837 | 3.62 | 1.50 | 1.14 | 1.68 | 1.46 | 1.30 | 0.72 | 0.78 | 0.66 | |
| 1838 | 1.64 | 0.75 | 0.76 | 1.32 | 1.40 | 1.67 | 0.82 | 1.40 | 3.84 | |
| 1839 | 0.70 | 0.80 | 0.58 | 4.06 | 2.98 | 0.94 | 1.08 | 3.54 | 0.70 | |

| | | | | | | | | | |
|---|---|---|---|---|---|---|---|---|---|
| 1840 | 1.68 | 2.20 | 1.54 | 2.12 | 1.16 | 1.08 | 1.40 | 2.72 | 1.28 |
| 1841 | 1.44 | 1.12 | 1.32 | 1.64 | 0.90 | 0.75 | 0.64 | 2.82 | 2.78 |
| 1842 | 0.54 | 1.22 | 1.16 | 0.64 | 0.47 | 2.10 | 0.68 | 1.44 | 0.96 |
| 1843 | 1.60 | 1.64 | 2.50 | 1.34 | 0.34 | 1.04 | 1.98 | 2.58 | 0.52 |
| 1844 | 4.14 | | 2.06 | 0.24 | 0.58 | 0.78 | 0.86 | 1.34 | 1.76 |
| 1845 | 2.42 | 1.70 | 1.14 | 0.70 | 1.02 | 1.03 | 1.20 | 1.66 | 0.88 |
| 1846 | 1.54 | | 2.46 | 1.16 | 1.18 | 0.82 | 1.46 | 0.49 | 0.56 |
| 1847 | 1.18 | 2.74 | 1.66 | 1.12 | 0.84 | 1.28 | 0.56 | 1.86 | 2.16 |
| 1848 | 1.44 | 1.56 | 2.68 | 0.68 | 2.28 | 1.00 | 0.72 | 1.24 | 1.48 |
| 1849 | 1.36 | 0.40 | 2.30 | 0.92 | 1.28 | 0.72 | 1.52 | 2.08 | 1.12 |
| 1850 | 2.56 | 1.92 | 1.84 | 2.68 | 2.80 | 1.20 | 1.20 | 3.68 | 6.88 |
| 1851 | 0.80 | 1.84 | 0.56 | 3.60 | 1.92 | 1.12 | 0.96 | 0.32 | 1.15 |
| 1852 | 1.06 | 0.88 | 1.15 | 4.38 | 1.47 | 1.69 | 0.66 | 4.16 | 1.19 |
| 1853 | 0.92 | 1.33 | 1.03 | 1.12 | 2.39 | 0.42 | 1.03 | 2.36 | 2.14 |
| 1854 | 0.83 | 1.60 | 1.25 | 1.88 | 2.57 | 1.50 | 1.58 | 0.48 | 2.33 |
| 1855 | 3.37 | 3.08 | 0.80 | 1.33 | 0.39 | 1.23 | 1.93 | 0.75 | 0.70 |
| 1856 | | 1.30 | 0.63 | 1.97 | 2.93 | 0.66 | 1.30 | 4.23 | 2.42 |
| 1857 | 1.50 | 0.54 | 1.55 | 3.68 | 1.28 | 0.96 | 2.43 | 2.00 | 0.87 |
| 1858 | 1.12 | 1.18 | 0.35 | 1.28 | 1.00 | 3.86 | 1.35 | 2.21 | 1.64 |

[58] The following table shows the record of rain-fall, as kept for one year; it was selected as a representative year, the total quantity falling being equal to the average. For the year 1840: Number of rainy days, 50; total rain-fall, 42.00.

| Days | January 1840 | February | March | April | May | June | July | August | September | O |
|---|---|---|---|---|---|---|---|---|---|---|
| 1 | | | | 0.55 | 0.14 | | | 2.72 | | |
| 2 | | | | | | | | 0.08 | | |
| 3 | | | 0.32 | | | | | | | |
| 4 | | | | | | 1.08 | 0.10 | | | |
| 5 | | | | | 1.16 | | | | 0.63 | |
| 6 | | | | | | | | 0.50 | | |
| 7 | | | | | | | | | | |
| 8 | | | | | | 0.20 | | | | |
| 9 | | | | | | | 0.25 | | | |
| 10 | | 2.20 | | | | | | | 1.28 | |
| 11 | | | | | | | | 0.10 | | |
| 12 | | | | 2.12 | | | | | | |
| 13 | | | | | | 0.14 | | | | |
| 14 | | 0.58 | | | | | | 0.70 | | |
| 15 | | | | | | | | | | |
| 16 | | | | | | | | | | |
| 17 | | | | | | | | | | |
| 18 | | | | | | | | | | |
| 19 | | | | | | 0.82 | 0.24 | | 0.68 | |
| 20 | | | 1.54 | | | | | | | |

| | | | | | | | | | |
|---|---|---|---|---|---|---|---|---|---|
| 21 | | | | | 0.98 | | | | |
| 22 | | | | 0.52 | | | | | |
| 23 | 1.68 | | | | | | | 0.96 | |
| 24 | | | | | | | 1.40 | | |
| 25 | | | | | | | | 0.16 | |
| 26 | | | | 0.18 | | | | | |
| 27 | | | | | | 0.17 | | | 0.30 |
| 28 | | | | | | | | | |
| 29 | | | | 1.80 | | | 0.10 | | |
| 30 | | | 1.42 | | | | | | |
| 31 | | | | | | | | | |
| Total | 1.68 | 2.78 | 3.28 | 5.17 | 2.28 | 2.41 | 2.09 | 5.22 | 2.89 |

[59] The average quantity of rain which has fallen in Waltham, during the important months of vegetation, from 1824 to 1858 inclusive — a period of thirty-five years — is for —

| April. | May. | June. | July. | Aug. | Sept. |
|---|---|---|---|---|---|
| 3.96 | 3.71 | 3.18 | 3.38 | 4.50 | 3.52 |

Average for the six months, 22.25.

It will be noticed, that the average for the month of August is about 33 per cent. larger than for June and July. The quantity of rain falling in each month, as registered at the Cambridge Observatory, is as follows:

MEAN OF OBSERVATIONS FOR TWELVE YEARS.

Jan. Feb. Mar. Apr. May. June. July. Aug. Sept. Oct. Nov. Dec.

2.39  3.19  3.47  3.64  3.74  3.13  2.57  5.47  4.27  3.73  4.57  4.31

| Spring. | Summer. | Autumn. | Winter. |
|---------|---------|---------|---------|
| 10.85   | 11.17   | 12.57   | 9.89    |

Average quantity per year, 44.48.

The quantity falling from January to July, is much less than falls from July to January.

The great quantity of snow which falls in New England during the Winter months, and is carried off mainly in the Spring, usually floods the low lands, and should be taken into account in establishing the size of pipe to be used in a system of drainage. The following observations of the average depth of snow, have been made at the places cited, and are copied, by Blodget, from various published notices:

| Oxford Co., Me. | 12 | years | 90 | inches | per year. |
|-----------------|----|-------|------|--------|-----------|
| Dover, N. H. | 10 | " | 68.6 | " | " |
| Montreal | 10 | " | 67 | " | " |
| Burlington, Vt. | 10 | " | 85 | " | " |
| Worcester, Mass. | 12 | " | 55 | " | " |
| Amherst, " | 7 | " | 54 | " | " |
| Hartford, Conn. | 24 | " | 43 | " | " |
| Lambertville, N. J. | 8 | " | 25.5 | " | " |
| Cincinnati | 16 | " | 19 | " | " |
| Burlington, Iowa | 4 | " | 15.5 | " | " |
| Beloit, Wisconsin | 3 | " | 25 | " | " |

[60] One-tenth the depth of snow is taken as its equivalent in water, for general purposes, though it gives too small a quantity of water in southern latitudes, and in extreme latitudes too great a quantity. The rule of reduction of snow to water, in cold climates, is one inch of water to twelve of snow.

The proportion of the annual downfall of rain which is collectable into reservoirs—or, in other words, the per-centage of the rain-fall which drains off—is well shown in a table used by Ellwood Morris, Esq., C. E., in an article on "The Proposed Improvement of the Ohio River" (Jour. Frank. Inst., Jan., 1858), in which we find, that, in eighteen series of observations in Great Britain, the ratio, or per cent. of the rain-fall which drains off is 65½, or nearly two-thirds the rain-fall.

Seven series of observations in America are cited as follows:

| No. | Name of Drainage Area. | Annual rain-fall, in inches. | Drainage flowing away, in inches. | Ratio, or per ct. of the rain which drains off. | Authorities. |
|---|---|---|---|---|---|
| 1 | Schuylkill Navigation Reservoirs | 36 | 18 | 50 | Morris and Smith. |
| 2 | Eaton Brook | 34 | 23 | 66 | McAlpine. |
| 3 | Madison Brook | 35 | 18 | 50 | McAlpine. |
| 4 | Patroon's Brook | 46 | 25 | 55 | McAlpine. |
| 5 | Patroon's Brook | 42 | 18 | 42 | McAlpine. |
| 6 | Long Pond | 40 | 18 | 44 | Boston Water Com'rs. |
| 7 | West Fork Reservoir | 36 | 14 | 40 | W. Milnor Roberts. |

| Totals | 269 | 134 | 347 |
| Averages | 38 | 19 | 50 |

These examples show an average rain-fall of thirty-eight vertical inches, and an annual amount, collectable in reservoirs, of nineteen inches, or fifty per cent.

The per-centage of water of drainage from land under-drained [61] with tile, would be greater than that which is collectable in reservoirs from ordinary gathering-grounds.

If a soil were perfectly saturated with water, that is, held as much water in suspension as possible to hold without draining off, and drains were laid at a proper depth from the surface, and in sufficient number to take off all surplus water, then the entire rain-fall upon the surface would be water of drainage—presuming, of course, the land to be level, and the air at saturation, so as to prevent evaporation. The water coming upon the surface, would force out an equal quantity of water at the bottom, through the drains—the time occupied by the process, varying according to the porous or retentive nature of the soil; but in ordinary circumstances, it would be, perhaps, about forty-eight hours. Drains usually run much longer than this after a heavy rain, and, in fact, many run constantly through the year, but they are supplied from lands at a higher level, either near by or at a distance.

If, on the other hand, the soil were perfectly dry, holding no water in suspension, then there would be no water of drainage until the soil had become saturated.

Evaporation is constantly carrying off great quantities of water during the warm months, so that under-drained soil is seldom in the condition of saturation, and, on account of the supply by capillary attraction and by dew, is never thoroughly dry; but the same soil will, at different times, be at various points between saturation and dryness, and the water of drainage will be consequently a greater or less per centage of the rain-fall.

An experiment made by the writer, to ascertain what quantity of water a dry soil would hold in suspension, resulted as follows: A soil was selected of about average porosity, so that the result might be, as nearly as possible, a mean for the various kinds of soil, and dried by several days' baking. The quantity of soil then being carefully [62] measured, a measured quantity of water was supplied slowly, until it began to escape at the bottom. The quantity draining away was measured and deducted from the total quantity supplied. It was thus ascertained that one cubic foot of earth held 0.4826+ cubic feet of water, which is a little more than three and one-half gallons. A dry soil, four feet deep, would hold a body of water equal to a rain-fall of 23.17 inches, vertical depth, which is more than would fall in six months.

The quantity which is not drained away is used for vegetation or evaporated; and the fact, that the water of drainage is so much greater in proportion to the rain-fall in England than in this country, is owing to the humidity of that climate, in which the evaporation is only about half what it is in this country.

The evaporation from a reservoir surface at Baltimore, during the Summer months, was assumed by Colonel Abert to be to the quantity of rain as two to one.

Dr. Holyoke assigns the annual quantity evaporated at Salem, Mass., at fifty-six inches; and Colonel Abert quotes several authorities at Cambridge, Mass., stating the quantity at fifty-six inches. These facts are given by Mr. Blodget, and also the table below.

QUANTITY OF WATER EVAPORATED, IN INCHES, VERTICAL DE|

| | Jan. | Feb. | Mar. | Apr. | May. | June. | July. | Aug. | Sept. | Oct. | Nov. | |
|---|---|---|---|---|---|---|---|---|---|---|---|---|
| Whitehaven, England, mean of 6 years | 0.88 | 1.04 | 1.77 | 2.54 | 4.15 | 4.54 | 4.20 | 3.40 | 3.12 | 1.93 | 1.32 | |
| Ogdensburg, N. Y., 1 yr. | 1.65 | 0.82 | 2.07 | 1.63 | 7.10 | 6.74 | 7.79 | 5.41 | 7.40 | 3.95 | 3.66 | |
| Syracuse, N. Y., 1 year | 0.67 | 1.48 | 2.24 | 3.42 | 7.31 | 7.60 | 9.08 | 6.85 | 5.33 | 3.02 | 1.33 | |

The quantity for Whitehaven, England, is reported by J. F. Miller. It was very carefully observed, from 1843 to 1848—the evaporation being from a copper vessel, protected from rain. The district is one of the wettest of England—the mean quantity of rain, for the same time, having been 45.25 inches.

This shows a great difference in the capacity of the air [63] to absorb moisture in England and the United States; and as evaporation is a cooling process, there is greater necessity for under-draining in this country than in England, supposing circumstances in other respects to be similar.

Evaporation takes place at any point of temperature from 32°, or lower, to 212°—at which water boils. It is increased by heat, but is not caused solely by it—for a north-west wind in New-England evaporates water, and dries the earth more rapidly than the heat alone of a Summer's day; and when, under ordinary circumstances, evaporation from a water-surface is slow, it becomes quite active when brought in close proximity to sulphuric acid, or other vapor-absorbing bodies.

The cold which follows evaporation is caused by a loss of the heat which is required for evaporation, and which passes off with the vapor, as a solution, in the atmosphere; and as heat leaves the body to aid evaporation, it is evident that that body cannot be cooled by the process, below the dew-point at which evaporation ceases. The popular notion that a body may be cooled almost to the freezing-point, in a hot Summer day, by the action of heat alone, is, then, erroneous. But still, the amount of heat which is used up in evaporating stagnant water from undrained land, that might otherwise go towards warming the land and the roots of crops, is a very serious loss.

The difference in the temperature of a body, resulting from evaporation, may reach 25° in the desert interior of the American continent; but, in the Eastern States, it is not often more than 15°.

The temperature of evaporation is the reading of a wet-bulb-thermometer (the bulb being covered with moistened gauze) exposed to the natural evaporation; and the difference between that reading and the reading of a dry-thermometer, is the expression of the cold resulting from evaporation. [64]

When the air is nearly saturated, the temperature of the air rarely goes above 74°; but, if so, the moisture in the air prevents the passing away of insensible perspiration, and the joint action of heat and humidity exhausts the vital powers, causing sun-stroke, as it is called. At New York city, August 12th to 14th, 1853, the wet-thermometer stood at 80° to 84°; the air, at 90° to 94°. The mortality, from this joint effect, was very great—over two hundred persons losing their lives in the two days, in that city.

From very careful observations, made by Lorin Blodget, in 1853, at Washington, it was found that the difference between the wet and dry thermometer was 18½° at 4 P. M., June 30th, and 16° at 2 P. M. on July 1st—the temperature of the air being 98° on the first day, and 95° on the second; but such excesses are unusual.

The following table has been compiled from Mr. Blodget's notice of the peculiarities of the Summer of 1853:

The dates are such as were selected to illustrate the extreme temperatures of the month, and the degrees represent the differences between the wet and dry thermometer. The observations were made at 3 P. M.:

| Locality. | Dates. | | Differences. | | |
|---|---|---|---|---|---|
| | June, 1853. | | | | |
| Burlington, Vt. | 14th to 30th | ranged from | 8° | to | 17° |
| Montreal | 14th to 30th | " | 6 | to | 17 |
| Poultney, Iowa | 10th to 30th | " | 9 | to | 16 |
| Washington | 20th to 30th | " | 8.5 | to | 16 |
| Baltimore | 13th to 30th | " | 7.4 | to | 20.2 |
| Savannah | 13th to 30th | " | 5.2 | to | 17.3 |
| Austin, Texas | 10th to 30th | " | 4 | to | 24 |
| Clarkesville, Tenn. | 4th to 30th | " | 10.3 | to | 20.5 |

August.

| | | | |
|---|---|---|---|
| Bloomfield, N. J. | 9th to 14th | " | 5 to 15 |
| Austin, Texas | 6th to 12th | " | 0 to 19 |
| Philadelphia | 10th to 15th | " | 8 to 14 |
| Jacksonville, Fla. | 10th to 15th | " | 6 to 8 |

[65] Observations by Lieut. Gillis, at Washington, give mean differences between wet and dry thermometers, from March, 1841, to June, 1842, as follows:

Observations at 3 P. M.:

**Jan. Feb. Mar. Apr. May. June. July. Aug. Sept. Oct. Nov. Dec.**

3°.08 4°.40 6°.47 5°.37 7°.05 8°.03 8°.89 5°.29 5°.63 4°.61 4°.77 2°.03

A mean of observations for twenty-five years at the Radcliffe Observatory, Oxford, England, gives a difference between the wet and dry thermometer equal to about two-thirds the difference, as observed by Lieutenant Gillis, at Washington.

On the 12th day of August, 1853, in Austin, Texas, the air was perfectly saturated at a temperature of 76°, which was the dew-point, or point of the thermometer at which dew began to form. The dew-point varies according to the temperature and the humidity of the atmosphere; it is usually a few degrees lower than the temperature of evaporation—never higher.

From observations made at Girard College, by Prof. A. D. Bache, in the years 1840 to 1845, we find, that for April, 1844, the dew-point ranged from 4° to 16° lower than the temperature of the air; in May, from 4° to 14° lower; in June, from 6° to 20° lower; in July, from 4° to 17°; in August, from 6° to 15° lower; and in September, from 6° to 21° lower. The dew-point is, then, during the important months of vegetation, within about 20° of the temperature of the air. The temperature of the dew-point, as observed by Prof. Bache, was highest in August, 1843, being 66°, and lowest in January, 1844, being 18°; in July, 1844, it was 64°, and in February, 1845, it was 25°. Its hourly changes during each day are quite marked, and follow, with some degree of regularity, the changes in the temperature of the air; their greatest departure from each other being at the hottest hour of the day, which is two or three hours after noon, and the least at the coldest [66] hour which is four or five hours after midnight. The average temperature of the dew-point in April, May, and June, 1844, was, at midnight, 50½°, air, 57°; five hours after midnight, dew-point, 49°, air 54°; three hours after noon, dew-point, 54°, air, 63½°. The average temperature for July, August and September, was, at midnight, dew-point, 58½°, air, 65°; five hours after midnight, dew-point, 58°, air, 62°; three hours after noon, dew-point, 60½°, air, 78°. The average temperature for the year was, at midnight, dew-point, 42°, air, 48°; five hours after midnight, dew-point, 41°, air, 46°; three hours after noon, dew-point, 44½°, air, 59°.

The relative humidity of the atmosphere, or the amount of vapor held in suspension in the air, in proportion to the amount which it might hold, was, in the year 1858, as given in the journal of the Franklin Institute, for

|           | **Philadelphia.** |          | **Somerset Co.** |          |
|-----------|-------------------|----------|------------------|----------|
| April     | 49                | per cent.| —                | 2 P. M.  |
| May       | 59                | "        | 72               | "        |
| June      | 55                | "        | 63               | "        |
| July      | 50                | "        | 61               | "        |
| August    | 55                | "        | 58               | "        |
| September | 50                | "        | 57               | "        |

The saturation often falls to 30 per cent., but with great variability. Evaporation goes on most rapidly when the per centage of saturation is lowest; and, as before observed, the cause of the excess of evaporation in this country over that of England is the excessive humidity of that climate and the dryness of this. It has also been said that there is greater need for drainage in the United States on this account; and, as the warmth induced by draining is somewhat, in its effect, a merchantable product, it may be well to consider it for a moment in that light.

First: The drained land comes into condition for [67] working, a week or ten days earlier in the Spring than other lands.

Secondly: The growth of the crops is quickened all through the Summer by an increase of several degrees in the temperature of the soil.

Thirdly: The injurious effects of frost are kept off several days later in the Fall.

Of the value of these conditions, the farmer, who has lost his crops for lack of a few more warm days, may make his own estimates. In Roxbury, Mr. I. P. Rand heats up a portion of his land, for the purpose of raising early plants for the market, by means of hot water carried by iron pipes under the surface of the ground. In this manner he heats an area equal to 100 feet by 12 feet, by burning about one ton of coal a month. The increase of temperature which, in this case, is caused by that amount of coal, can, in the absence of direct measurement, only be estimated; but it, probably, will average about 30°, day and night, throughout the month. In an acre the area is 36.4 times as great as that heated by one ton of coal; the cost being in direct proportion to the area, 36.4 tons of coal would be required to heat an acre; which, at $6 per ton, would cost $217.40. To heat an acre through 10°, would cost, then, $72.47. It may be of interest to consider how much coal would be required to evaporate from an undrained field that amount of water which might be carried off by under-drains, but which, without them, is evaporated from the surface. It may be taken as an approximate estimate, that the evaporation from the surface of an undrained retentive field, is equal to two inches vertical depth of water for each of the months of May, June, July, and August; which is equal to fifty-four thousand three hundred and five gallons, or eight hundred and sixty-two hogsheads per acre for each month. If this quantity of water were evaporated by means of a coal fire, [68] about 22⅔ tons of coal would be consumed, which, at $6 a ton, would cost $136. The cost of evaporating the amount of water which would pass off in one day from an acre would be about $4.53. It is probable that about half as much water would be evaporated from thorough-drained land, though, by some experiments, the proportion has been made greater — in which case the loss of heat resulting from an excess of moisture evaporated from undrained retentive land, over that which would be evaporated from drained land, would be equal to that gained by 11⅓ tons of coal, which would cost $68; and this for each acre, in each of the three months. At whatever temperature a liquid vaporizes, it absorbs the same total quantity of heat.

The latent heat of watery vapor at 212° is 972°; that is, when water at 212° is converted into vapor at the same temperature, the amount of heat expended in the process is 972°. This heat becomes latent, or insensible to the thermometer. The heat rendered latent by converting ice into water is about 140°. There are 7.4805 gallons in a cubic foot of water which weighs 62.38 lbs."

We have seen that a sea of water, more than three feet deep over the whole face of the land, falls annually from the clouds, equal to 4,000 tons in weight to every acre. We would use enough of this water to dissolve the elements of fertility in the soil, and fit them for the food of plants. We would retain it all in our fields, long enough to take from it its stores of fertilizing substances, brought from reeking marshes and steaming cities on cloud-wings to our farms. We would, after taking enough of its moisture to cool the parched earth, and to fit the soil for germination and vegetable growth, discharge the surplus, which must otherwise stagnate in the subsoil, by rapid drainage into the natural streams and rivers.

Evaporation proceeds more rapidly from a surface of [69] water, than from a surface of land, unless it be a saturated surface. It proceeds more rapidly in the sun than in the shade, and it proceeds again more rapidly in warm than in cold weather. It varies much with the culture of the field, whether in grass, or tillage, or fallow, and with its condition, as to being dry or wet, and with its formation, whether level or hilly. Yet, with all these variations, very great reliance may be placed upon the ascertained results of the observations already at our command.

We have seen that evaporation from a water surface is, in general, greater than from land, and here we may observe one of those grand compensating designs of Providence which exist through all nature.

If the same quantity of water fell upon the sea and the land, and the evaporation were the same from both, then all the rivers running into the sea would soon convey to it all the water, and the sea would be full. But though nearly as much water falls on the sea as on the land, yet evaporation is much greater from the water than from land.

About three feet of rain falls upon the *water*, while the evaporation from a water surface far exceeds that amount. In the neighborhood of Boston, evaporation from water surface is said to be 56 inches in the year, and in the State of New York, about 50 inches; while, in England, it is put by Mr. Dalton at 44.43 inches, and, by others, much lower.

Again, about three feet of water annually falls upon the *land*, while the evaporation from the land is but little more than 20 inches. If this water fell upon a flat surface of soil, with an impervious subsoil of rock or clay, we should have some sixteen inches of water in the course of the year more than evaporates from the land. If a given field be dish-shaped, so as to retain it all, it must become a pond, and so remain, except in Summer, [70] when greater evaporation from a water surface may reduce it to a swamp or marsh.

With 16 or 18 inches more water falling annually on all our cultivated fields than goes off by evaporation, is it not wise to inquire by what process of Nature or art this vast surplus shall escape?

Experiments have been made with a view to determine the proportion of evaporation and filtration, upon well-drained land, in different months. From an able article in the N. Y. Agricultural Society for 1854, by George Geddes, we copy the following statement of valuable observations upon these points.

It will be observed that, in the different observations collected in this chapter, results are somewhat various. They have been brought together for comparison, and will be found sufficiently uniform for all practical purposes in the matter of drainage.

"The experiments upon evaporation and drainage, made on Mr. Dalton's plan, were in vessels three feet deep, filled with soil just in the condition to secure perfect freedom from excess of water, and the drainage was determined by the amount of water that passed out of the tube at the bottom. These experiments have been most perfectly made in England by Mr. John Dickinson. The following table exhibits the mean of eight years:

| | October to March. | | | April to September. | | | Total each ye | | |
|---|---|---|---|---|---|---|---|---|---|
| Year. | Rain. | Filtration | Per cent filtered. | Rain. | Filtration | Per cent filtered. | Rain. | Filtration | P fi |
| 1836 | 18.80 | 15.55 | 82.7 | 12.20 | 2.10 | 17.3 | 31.00 | 17.65 | |
| 1837 | 11.30 | 6.85 | 60.6 | 9.80 | 0.10 | 1.0 | 21.10 | 6.95 | |
| 1838 | 12.32 | 8.45 | 68.8 | 10.81 | 0.12 | 1.2 | 23.13 | 8.57 | |
| 1839 | 13.87 | 12.31 | 88.2 | 17.41 | 2.60 | 15.0 | 31.28 | 14.91 | |
| 1840 | 11.76 | 8.19 | 69.6 | 9.68 | 0.00 | 0.0 | 21.44 | 8.19 | |
| 1841 | 16.84 | 14.19 | 84.2 | 15.26 | 0.00 | 0.0 | 32.10 | 14.19 | |
| 1842 | 14.28 | 10.46 | 73.2 | 12.15 | 1.30 | 10.7 | 26.43 | 11.76 | |
| 1843 | 12.43 | 7.11 | 57.2 | 14.04 | 0.99 | 7.1 | 26.47 | 8.10 | |
| Mean | 13.95 | 10.39 | 74.5 | 12.67 | 0.90 | 7.1 | 26.61 | 11.29 | |

[71] "A soil that holds no water for the use of plants below six inches, will suffer from drouth in ten days in June, July, or August. If the soil is in suitable condition to hold water to the depth of three feet, it would supply sufficient moisture for the whole months of June, July, and August.

"M. de la Hire has shown that, at Paris, a vessel, sixteen inches deep, filled with sand and loam, discharged water through the pipe at the bottom until the 'herbs' were somewhat grown, when the discharge ceased, and the rains were insufficient, and it was necessary to water them. The fall of water at Paris is stated, in this account, at twenty inches in the year, which is less than the average, and the experiment must have been made in a very dry season; but the important point proved by it is, that the plants, when grown up, draw largely from the ground, and thereby much increase the evaporation from a given surface of earth. The result of the experiment is entirely in accordance with what would have been expected by a person conversant with the laws of vegetation.

"The mean of each month for the eight years is:

| Months. | Rain. | Filtration. | Per cent filtered. |
|---|---|---|---|
| | Inches. | Inches. | |
| January | 1.84 | 1.30 | 70.7 |
| February | 1.79 | 1.54 | 78.4 |
| March | 1.61 | 1.08 | 66.6 |
| April | 1.45 | 0.30 | 21.0 |
| May | 1.85 | 0.11 | 5.8 |
| June | 2.21 | 0.04 | 1.7 |
| July | 2.28 | 0.04 | 1.8 |
| August | 2.42 | 0.03 | 1.4 |
| September | 2.64 | 0.37 | 13.9 |
| October | 2.82 | 1.40 | 49.5 |
| November | 3.83 | 3.26 | 84.9 |
| December | 1.64 | 1.80 | 110.0 |

"The filtration from April to September is very small—practically nothing; but during those months we have 12.67 inches of rain—that is, we have two inches a month for evaporation besides the quantity in the earth on the first day of April. From October to March we have 10.39 inches filtered out of 13.95 inches, the whole fall. 'Of this Winter portion of 10.39, we must allow at least six inches for floods running away at the time of the rain, and then we have only 4.39 inches left for the supply of rivers and wells.' (Breadmore, p. 34.) [72]

"It is calculated in England that the ordinary Summer run of streams does not exceed ten cubic feet per minute per square mile, and that the average for the whole year, due to springs and ordinary rains, is twenty feet per minute per square mile, exclusive of floods—and assuming no very wet or high mountain districts (Breadmore, p. 34)—which is equal to about four inches over the whole surface. If we add to this the six inches that are supposed to run off in freshets, we have ten inches discharged in the course of the year by the streams. The whole filtration was 11.29 inches—10.39 in the Winter, and .90 in the Summer. The remainder, 1.29 inches, is supposed to be consumed by wells and excessive evaporation from marshes and pools, from which the discharge is obstructed, by animals, and in various other ways. These calculations were made from experiments running through eight years, in which the average fall of water was only 26.61 inches per annum. When the results derived from them are applied to our average fall of 35.28 inches, we have for the water that constitutes the Summer flow of our streams 13.25 cubic feet per minute per mile of the country drained, and for the average annual flow, exclusive of freshets, 26.50 cubic feet per mile per minute. That is to say, of the 35.28 inches of water that fall in the course of the year, 5.30 run away in the streams as the average annual flow, 7.95 run away in the freshets, and 20.47 evaporate from the earth's surface, leaving 1.56 for consumption in various ways. In the whole year the drainage is nearly equal to one cubic foot per second per square mile (.976), no allowance being made for the 1.56 inches which is lost as before stated. These calculations are based upon English experiments. Mr. McAlpine, late State engineer and surveyor, in making his calculations for supplying the city of Albany with water (page 22 of his Report to the Water Commissioners), takes 45 per cent of the fall as available for the use of the city. Mr. Henry Tracy, in his Report to the Canal Board of 1849 (page 17), gives the results of the investigations in the valleys of Madison Brook, in Madison County, and of Long Pond, near Boston, Mass., as follows:

| Year. | Name of | Fall of | Water ran | Evaporation | Ratio of |
|---|---|---|---|---|---|

| | valley. | rain and snow in valley. | off in inches. | from surface of ground. | drainage. |
|---|---|---|---|---|---|
| 1835 | Madison Brook | 35.26 | 15.83 | 19.43 | 0.449 |
| 1837 | Long Pond | 26.65 | 11.70 | 14.95 | 0.439 |
| 1838 | Do | 38.11 | 16.62 | 21.49 | 0.436 |
| Mean | | | | | 0.441 |

[73] "Madison Brook drains 6,000 acres, and Long Pond 11,400 acres. Mr. Tracy makes the following comment on this table: 'It appears that the evaporation from the surface of the ground in the valley of Long Pond was about 44 per cent more in 1838 than it was in 1837, while the ratio of the drainage differed less than one per cent the same years.'

"Dr. Hale states the evaporation from water-surface at Boston to be 56 inches in a year. (Senate Doc., No. 70, for 1853.)

"The following table contains the results arrived at by Mr. Coffin, at Ogdensburgh, and Mr. Conkey, at Syracuse, in regard to the evaporation from water-surface:

| Months. | Coffin, at Ogdensburgh, in 1838. | | Conkey, at Syracuse, in 1852. | |
|---|---|---|---|---|
| | Rain. | Evaporation. | Rain. | Evaporation. |
| January | 2.36 | 1.652 | 3.673 | 0.665 |
| February | 0.97 | 0.817 | 1.307 | 1.489 |
| March | 1.18 | 2.067 | 3.234 | 2.239 |
| April | 0.40 | 1.625 | 3.524 | 3.421 |
| May | 4.81 | 7.100 | 4.491 | 7.309 |
| June | 3.57 | 6.745 | 3.773 | 7.600 |

| | | | | |
|---|---|---|---|---|
| July | 1.88 | 7.788 | 2.887 | 9.079 |
| August | 2.55 | 5.415 | 2.724 | 6.854 |
| September | 1.01 | 7.400 | 2.774 | 5.334 |
| October | 2.73 | 3.948 | 4.620 | 3.022 |
| November | 2.07 | 3.659 | 4.354 | 1.325 |
| December | 1.08 | 1.146 | 4.112 | 1.863 |
| Total | 24.61 | 49.362 | 41.473 | 50.200 |

"The annual fall of water in England, is stated, by Mr. Dalton, to be 32 inches. In this State, it is 35.28 inches. The evaporation from water-surface in England, is put, by Mr. Dalton, at 44.43 inches. The fall is less, and the evaporation is less, in England than here; and the fall, in each case, bears the same proportion to the evaporation, very nearly; and it appears that the experiments made on the two sides of the ocean, result in giving very nearly the same per centage of drainage. In England, it is 42.4 per cent.; in this State, it is 44.1. In England, the experiments were made on a limited scale compared with ours; but the results agree so well, that great confidence may safely be placed in them."

In reviewing the whole subject of rain, and of evaporation [74] and filtration, we seem to have evidence to justify the opinion, that with considerable more rain in this country than in England, and with a greater evaporation, because of a clearer sky and greater heat, we have a larger quantity of surplus water to be disposed of by drainage.

The occasion for thorough-drainage, however, is greater in the Northern part of the United States than in England, upon land of the same character; because, as we have already seen, rain falls far more regularly there than here, and never in such quantities in a single day; and because there the land is open to be worked by the plough nearly every day in the year, while here for several months our fields are locked up in frost, and our labor for the Spring crowded into a few days. There, the water which falls in Winter passes into the soil, and is drained off as it falls; while here, the snow accumulates to a great depth, and in thawing floods the land at once.

Both here and in England, much of the land requires no under-draining, as it has already a subsoil porous enough to allow free passage for all the surplus water; and it is no small part of the utility of understanding the principles of drainage, that it will enable farmers to discriminate—at a time when draining is somewhat of a fashionable operation with amateurs—between land that does and land that does *not* require so expensive an operation.

# CHAPTER IV [75]
# DRAINAGE OF HIGH LANDS—WHAT LANDS REQUIRE DRAINAGE.

What is High Land?—Accidents to Crops from Water.—Do Lands need Drainage in America?—Springs.—Theory of Moisture, with Illustrations.—Water of Pressure.—Legal Rights as to Draining our Neighbor's Wells and Land.—What Lands require Drainage?—Horace Greeley's Opinion.—Drainage more Necessary in America than in England; Indications of too much Moisture.—Will Drainage Pay?

By "high land," is meant land, the surface of which is not over-flowed, as distinguished from swamps, marshes, and the like low lands. How great a proportion of such lands would be benefitted by draining, it is impossible to estimate.

The Committee on Draining, in their Report to the State Agricultural Society of New York, in 1848, assert that, "There is not one farm out of every seventy-five in this State, but needs draining—yes, much draining—to bring it into high cultivation. Nay, we may venture to say, that every wheat-field would produce a larger and finer crop if properly drained." The committee further say: "It will be conceded, that no farmer ever raised a good crop of grain on wet ground, or on a field where pools of water become masses of ice in the Winter. In such cases, the grain plants are generally frozen out and perish; or, if any survive, they never arrive at maturity, nor produce a well-developed seed. In fact, every observing farmer knows that stagnant water, whether on the surface of his soil, or within reach of the roots of his plants, always does them injury." [76]

The late Mr. Delafield, one of the most distinguished agriculturists of New York, said in a public address:

"We all well know that wheat and other grains, as well as grasses, are never fully developed, and never produce good seed, when the roots are soaked in moisture. No man ever raised good wheat from a wet or moist subsoil. Now, the farms of this country, though at times during the Summer they appear dry, and crack open on the surface, are not, in fact, dry farms, for reasons already named. On the contrary, for nine months out of twelve, they are moist or wet; and we need no better evidence of the fact, than the annual freezing out of the plants, and consequent poverty of many crops."

If we listen to the answers of farmers, when asked as to the success of their labors, we shall be surprised, perhaps, to observe how much of their want of success is attributed to *accidents*, and how uniformly these accidents result from causes which thorough draining would remove. The wheat-crop of one would have been abundant, had it not been badly frozen out in the Fall; while another has lost nearly the whole of his, by a season too wet for his land. A farmer at the West has planted his corn early, and late rains have rotted the seed in the ground; while one at the East has been compelled, by the same rains, to wait so long before planting, that the season has been too short. Another has worked his *clayey* farm so wet, because he had not time to wait for it to dry, that it could not be properly tilled. And so their crops have wholly or partially failed, and all because of too much cold water in the soil. It would seem, by the remarks of those who till the earth, as if there were never a season just right — as if Providence had bidden us labor for bread, and yet sent down the rains of heaven so plentifully as always to blight our harvests. It is rare that we do not have a most remarkable season, with respect to moisture, especially. Our potatoes are rotted by the Summer showers, or cut off by a Summer drought; and when, as in the season of 1856, in [77] New England, they are neither seriously diseased nor dried up, we find at harvest-time that the promise has belied the fulfillment; that, after all the fine show above ground, the season has been too wet, and the crop is light. We frequently hear complaint that the season was too *cold* for Indian corn, and that the ears did not fill; or that a sharp drought, following a wet Spring, has cut short the crop. We hear no man say, that he lacked skill to cultivate his crop. Seldom does a farmer attribute his failure to the poverty of his soil. He has planted and cultivated in such a way, that, in a *favorable season*, he would have reaped a fair reward for his toil; but the season has been too wet or too dry; and, with full faith that farming will pay in the long run, he resolves to plant the same land in the same manner, hoping in future for better luck.

*Too much cold water* is at the bottom of most of these complaints of unpropitious seasons, as well as of most of our soils; and it is in our power to remove the cause of these complaints and of our want of success.

> "The fault, dear Brutus, is not in our stars,
> But in ourselves."

We must underdrain all the land we cultivate, that Nature has not already underdrained, and we shall cease complaints of the seasons. The advice of Cromwell to his soldiers: "Trust God, and keep your powder dry," affords a good lesson of faith and works to the farmer. We shall seldom have a season, upon properly drained land, that is too wet, or too cold, or even too dry; for thorough draining is almost as sure a remedy for a drought, as for a flood.

*Do lands need under draining in America?* It is a common error to suppose that, because the sun shines more brightly upon this country than upon England, and because almost every Summer brings such a drought here as is unknown there, her system of thorough drainage can [78] have no place in agriculture on this side of the Atlantic. It is true that we have a clearer sky and a drier climate than are experienced in England; but it is also true that, although we have a far less number of showers and of rainy days, we have a greater quantity of rain in the year.

The necessity of drainage, however, does not depend so much upon the quantity of water which falls or flows upon land, nor upon the power of the sun to carry it off by evaporation, as upon *the character of the subsoil*. The vast quantity of water which Nature pours upon every acre of soil annually, were it all to be removed by evaporation alone, would render the whole country barren; but Nature herself has kindly done the work of draining upon a large proportion of our land, so that only a healthful proportion of the water which falls on the earth, passes off at the surface by the influence of the sun.

If the subsoil is of sand or gravel, or of other porous earth, that portion of the water not evaporated, passes off below by natural

drainage. If the subsoil be of clay, rock, or other impervious substances, the downward course of the water is checked, and it remains stagnant, or bursts out upon the surface in the form of
springs.

As the primary object of drainage is to remove surplus water, it
may be well to consider with some care

## THE SOURCES OF MOISTURE.

*Springs.* — These are, as has been suggested, merely the water of
rain and snow, impeded in its downward percolation, and collected
and poured forth in a perennial flow at a lower level.

The water which falls in the form of rain and snow upon the soil
of the whole territory of the United States, east of the Rocky Mountains, each year, is sufficient to cover it to the depth of more than 3
feet. It comes upon the [79] earth, not daily in gentle dews to water
the plants, but at long, unequal intervals, often in storms, tempests,
and showers, pouring out, sometimes, in a single day, more than
usually falls in a whole month.

What becomes of all this moisture, is an inquiry especially interesting to the agriculturist, upon whose fruitful fields this flood of
water annually descends, and whose labor in seed-time would be
destroyed by a single Summer shower, were not Nature more
thoughtful than he, of his welfare. Of the water which thus falls
upon cultivated fields, a part runs away into the streams, either
upon the surface, or by percolation through the soil; a part is taken
up into the air by evaporation, while a very small proportion enters
into the constitution of vegetation. The proportion which passes off
by percolation varies according to the nature of the soil in the locality where it falls.

Usually, we find the crust of the earth in our cultivated fields, in
strata, or layers: first, a surface-soil of a few inches of a loamy nature, in which clay or sand predominates; and then, it may be, a
layer of sand or gravel, freely admitting the passage of water; and,
perhaps, next, and within two or three feet of the surface, a stratum
of clay, or of sand or gravel cemented with some oxyd of iron,
through which water passes very slowly, or not at all. These strata
are sometimes regular, extending at an equal depth over large

tracts, and having a uniform dip, or inclination. Oftener, however, in hilly regions especially, they are quite irregular—the impervious stratum frequently having depressions of greater or less extent, and holding water, like a bowl. Not unfrequently, as we cut a ditch upon a declivity, we find that the dip of the strata below has no correspondence with the visible surface of the field, but that the different strata lie nearly level, or are much broken, while the surface has a regular inclination. [80]

Underlying all soils, at greater or less depth, is found some bed of rock, or clay, impervious to water, usually at but few feet below the surface—the descending water meeting with obstacles to its regular descent. The tendency of the rain-water which falls upon the earth, is to sink directly downward by gravitation. Turned aside, however, by the many obstacles referred to, it often passes obliquely, or almost horizontally, through the soil. The drop which falls upon the hill-top sinks, perhaps, a few inches, meets with a bed of clay, glides along upon it for many days, and is at last borne out to be drunk up by the sun on some far-off slope; another, falling upon the sand-plain, sinks at once to the "water-line," or line of level water, which rests on clay beneath, and, slowly creeping along, helps to form a swamp or bog in the valley.

Sometimes, the rain which falls upon the high land is collected together by fissures in the rocks, or by seams or ruptures in the impervious strata below the surface, and finds vent in a gushing spring on the hill-side.

We feel confident that no better illustration of the theory of springs, as connected with our subject, can be found, than that of Mr. Girdwood, in the Cyclopedia of Agriculture—a work from which we quote the more liberally, because it is very expensive and rare in America:

"When rain falls on a tract of country, part of it flows over the surface, and makes its escape by the numerous natural and artificial courses which may exist, while another portion is absorbed by the soil and the porous strata which lie under it.

"Let the following diagram represent such a tract of country, and let the dark portions represent clay or other impervious strata, while the [81] lighter portions represent layers of gravel, sand, or chalk, permitting a free passage to water.

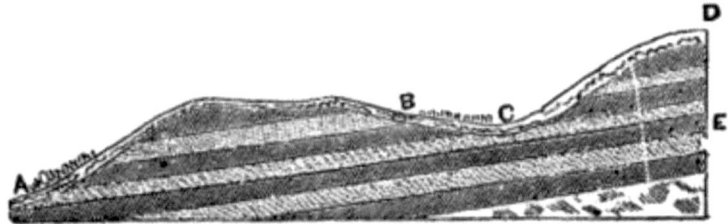

Fig. 5.

"When rain falls in such a district, after sinking through the surface-layer (represented in the diagram by a narrow band), it reaches the stratified layers beneath. Through these it still further sinks, if they are porous, until it reaches some impervious stratum, which arrests its directly-downward course, and compels it to find its way along its upper surface. Thus, the rain which falls on the space represented between B and D, is compelled, by the impervious strata, to flow towards C. Here it is at once absorbed, but is again immediately arrested by the impervious layer E; it is, therefore, compelled to pass through the porous stratum C, along the surface of E to A, where it pours forth in a fountain, or forms a morass or swamp, proportionate in size or extent to the tract of country between B and D, or the quantity of rain which falls upon it. In such a case as is here represented, it will be obvious that the spring may often be at a great distance from the district from which it derives its supplies; and this accounts for the fact, that drainage-works on a large scale sometimes materially lessen the supply of water at places remote from the scene of operations.

"In the instance given above, the water forming the spring is represented as gaining access to the porous stratum, at a point where it crops out from beneath an impervious one, and as passing along to its point of discharge at a considerable depth, and under several layers of various characters. Sometimes, in an undulating country, large tracts may rest immediately upon some highly-porous stratum — as from B to C, in the following diagram — rendering the necessity for draining less apparent; while the country from A to B, and from C to D, may be full of springs and marshes — arising, partly, from the rain itself, which falls in these latter districts, being unable to find a way of escape, and partly from the natural drainage of the more porous soils adjoining being discharged upon it.

Fig. 6.

"Again: the rocks lying under the surface are sometimes so full of fissures, that, although they themselves are impervious to water, yet, [82] so completely do these fissures carry off rain, that, in some parts of the county of Durham, they render the sinking of wells useless, and make it necessary for the farmers to drive their cattle many miles for water. It sometimes happens that these fissures, or cracks, penetrate to enormous depths, and are of great width, and filled with sand or clay. These are termed *faults* by miners; and some, which we lately examined, at distances of from three to four hundred yards from the surface, were from five to fifteen yards in width. These faults, when of clay, are generally the cause of springs appearing at the surface: they arrest the progress of the water in some of the porous strata, and compel it to find an exit, by passing to the surface between the clay and the faces of the ruptured strata. When the fault is of sand or gravel, the opposite effect takes place, if it communicates with any porous stratum; and water, which may have been flowing over the surface, on reaching it, is at once absorbed. In the following diagram, let us suppose that B represents such a clay-fault as has been described, and that A represents a sandy one, and that C and D represent porous strata charged with water. On the water reaching the fault at B, it will be compelled to find its way to the surface—there forming a spring, and rendering the retentive soil, from B to A, wet; but, as soon as it reaches the sandy-fault at A, it is immediately absorbed, and again reaches the porous strata, along which it had traveled before being forced to the surface at B. It will be observed, that the strata at the points of dislocation are not represented as in a line with the portions from which they have been dissevered. This is termed the upthrow of the fault, as at B; and the downthrow, as at A. For the sake of the illustration, the displacement is here shown as very slight; but, in some cases, these elevations and depressions of the strata extend to many hundreds of feet—as, for instance, at the mines of the British Iron Company, at Cefn-Mawre, in North Wales, where the downthrow of the fault is 360 feet.

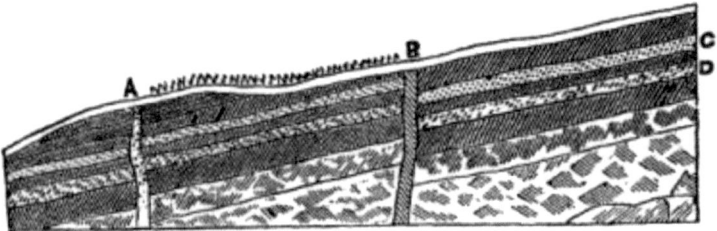

Fig. 7.

"Sometimes the strata are disposed in the form of a basin. In this case, the water percolating through the more elevated ground— near [83] what may be called the rim—collects in the lower parts of the strata towards the centre, there forcing its way to the surface, if the upper impervious beds be thin; or, if otherwise, remaining a concealed reservoir, ready to yield its supplies to the shaft or boring-rod of the well-sinker, and sometimes forming a living fountain capable of rising many feet above the surface. It is in this way that what are called Artesian wells are formed. The following diagram represents such a disposition of the strata as has just been referred to. The rain which falls on the tracts of country at A and B, gradually percolates towards the centre of the basin, where it may be made to give rise to an Artesian well, as at C, by boring through the superincumbent mass of clay; or it may force itself to the surface through the thinner part of the layer of clay, as at D—there forming a spring, or swamp.

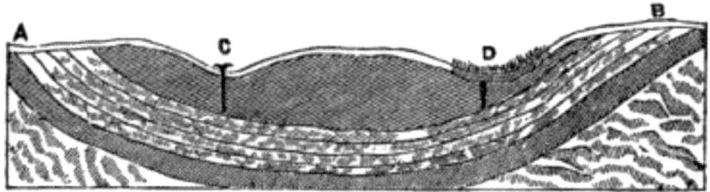

Fig. 8.

"Again: the higher parts of hilly ground are sometimes composed of very porous and absorbent strata, while the lower portions are more impervious—the soil and subsoil being of a very stiff and retentive description. In this case, the water collected by the porous layers is prevented from finding a ready exit, when it reaches the impervious layers, by the stiff surface-soil. The water is by this means dammed up in some measure, and acquires a considerable degree of pressure; and, forcing itself to the day at various places, it forms those extensive "weeping"-banks which have such an injurious effect upon many of our mountain-pastures. This was the form of spring, or swamp, to the removal of which Elkington principally turned his attention; and the following diagram, taken from a description of his system of draining, will explain the stratification and springs referred to, more clearly.

[84]

Fig. 9.

"In some districts, where clay forms the staple of the soil, a bed of sand or gravel, completely saturated with water, occurs at the depth of a few feet from the surface, following all the undulations of the country, and maintaining its position, in relation to the surface, over considerable tracts, here and there pouring forth its waters in a spring, or denoting its proximity, by the subaquatic nature of the herbage. Such a configuration is represented in the following diagram, where A represents the surface-soil; B, the impervious subsoil of clay; C, the bed of sandy-clay or gravel; and D, the lower bed of clay, resting upon the rocky strata beneath.

Fig. 10.

"Springs sometimes communicate with lakes or pools, at higher levels. In such cases, the quantity of water discharged is generally so great, as to form at once a brook or stream of some magnitude. These, therefore, hardly come under the ordinary cognizance of the land-drainer, and are, therefore, here merely referred to."

## THE WATER OF PRESSURE.

Water that issues from the land, either constantly, periodically, or even intermittently, may, perhaps, be properly termed a *spring*. But there is often much water in the soil which did not fall in rain upon that particular field, and which does not issue from it in any defined stream, but which is slowly passing through it by percolation from a higher source, to ooze out into some stream, or to pass off by evaporation; or, perhaps, farther on, to fall into crevices in the soil, and eventually form springs. As we find it in our field, it is neither rain-water, which has there fallen, nor spring-water, in any sense. It has been appropriately termed the *water of pressure*, to distinguish it from both rain and spring-water; and the recognition of this term will certainly be found convenient [85] to all who are engaged in the discussion of drainage.

The distinction is important in a legal point of view, as relating to the right of the land-owner to divert the sources of supply to mill-streams, or to adjacent lower lands. It often happens that an owner of land on a slope may desire to drain his field, while the adjacent owner below, may not only refuse to join in the drainage, but may believe that he derives an advantage from the surface-washing or the percolation from his higher neighbor. He may believe that, by deep drainage above, his land will be dried up and rendered worthless; or, he may desire to collect the water which thus percolates, into his land, and use it for irrigation, or for a water-ram, or for the supply of his barn-yard. May the upper owner legally proceed with the drainage of his own land, if he thus interfere with the interests of the man below?

Again: wherever drains have been opened, we already hear complaints of their effects upon wells. In our good town of Exeter, there seems to be a general impression on one street, that the drainage of a swamp, formerly owned by the author, has drawn down the wells on that street, situated many rods distant from the drains. Those wells are upon a sandy plain, with underlying clay, and the drains are cut down upon the clay, and into it, and may possibly draw off the water a foot or two lower through the whole village—if we can regard the water line running through it as the surface of a pond, and the swamp as a dam across its outlet.

The rights of land-owners, as to running water over their premises, have been fruitful of litigation, but are now well defined. In general, in the language of Judge Story,

"Every proprietor upon each bank of a river, is entitled to the land [86] covered with water in front of his bank to the middle thread of the stream, &c. In virtue of this ownership, he has a right to the use of the water flowing over it in its natural current, without diminution or obstruction. The consequence of this principle is, that no proprietor has a right to use the water to the prejudice of another. It is wholly immaterial whether the party be a proprietor above or below, in the course of the river, the right being common to all the proprietors *on* the river. No one has a right to diminish the quantity which will, according to the natural current, flow to the proprietor below, or to throw it back upon a proprietor above."

Chief Justice Richardson, of New Hampshire, thus briefly states the same position:

"In general, every man has a right to the use of the water flowing in a stream through his land, and if any one divert the water from its natural channel, or throw it back, so as to deprive him of the use of it, the law will give him redress. But one man may acquire, by grant, a right to throw the water back upon the land of another, and long usage may be evidence of such a grant. It is, however, well settled that a man acquires no such right by merely being the first to make use of the water."

We are not aware that it has ever been held by any court of law, or even asserted, that a land-owner may not intercept the percolating water in his soil for any purpose and at his pleasure; nor have we in mind any case in which the draining out of water from a well, by drainage for agricultural purposes, has subjected the owner of the land to compensation.

It is believed that a land-owner has the right to follow the rules of good husbandry in the drainage of his land, so far as the water of pressure is concerned, without responsibility for remote consequences to adjacent owners, to the owners of distant wells or springs that may be affected, or to mill-owners.

In considering the effect of drainage on streams and rivers, it appears that the results of such operations, so far as they can be appreciated, are, to lessen the value of water powers, by increasing the flow of water in times of [87] freshets, and lessening it in times of drought. It is supposed in this country, that clearing the land of timber has sensibly affected the value of "mill privileges," by increasing evaporation, and diminishing the streams. No mill-owner has been hardy enough to contend that a land-owner may not legally cut down his own timber, whatever the effect on the streams. So, we trust, no court will ever be found, which will restrict the land-owner in the highest culture of his soil, because his drainage may affect the capacity of a mill-stream to turn the water-wheels.

To return from our digression. It is necessary, in order to a correct apprehension of the work which our drains have to perform, to form a correct opinion as to how much of the surplus moisture in our field is due to each of the three causes to which we have referred — to wit, rain-water, which falls upon it; springs, which burst up from below; and water of pressure, stagnant in, or slowly percolating through it. The rain-tables will give us information as to the first; but as to the others, we must form our opinion from the structure of the earth around us, and observation upon the field itself, by its natural phenomena and by opening test-holes and experimental ditches. Having gained accurate knowledge of the sources of moisture, we may then be able to form a correct opinion whether our land requires drainage, and of the aid which Nature requires to carry off the surplus water.

## WHAT LANDS REQUIRE DRAINAGE?

The more one studies the subject of drainage, the less inclined will he be to deal in general statements. "Do you think it is profitable to underdrain land?" is a question a thousand times asked, and yet is a question that admits of no direct general answer. Is it profitable to fence land? is it profitable to plow land? are questions of much the same character. The answers to them all depend [88] upon circumstances. There is land that may be profitably drained, and fenced, and plowed, and there is a great deal that had better be let alone. Whether draining is profitable or not, depends on the value and character of the land in question, as well as on its condition as to water. Where good land is worth one hundred dollars an acre, it might be profitably drained; when, if the same land were worth but the Government price of $1.25 an acre, it might be better to make a new purchase in the neighborhood, than to expend ten times its value on a tract that cannot be worth the cost of the operation. Drainage is an expensive operation, requiring much labor and capital, and not to be thought of in a pioneer settlement by individual emigrants. It comes after clearing, after the building of log-houses and mills, and schoolhouses, and churches, and roads, when capital and labor are abundant, and when the good lands, nature-drained, have been all taken up.

And, again, whether drainage is profitable, depends not only on the value, but on the character of the soil as to productiveness when drained. There is much land that would be improved by drainage, that cannot be profitably drained. It would improve almost any land in New England to apply to it a hundred loads of stable manure to the acre; but whether such application would be profitable, must depend upon the returns to be derived from it. Horace Greeley, who has his perceptions of common affairs, and especially of all that relates to progress, wide awake, said, in an address at Peekskill, N. Y.:

"My deliberate judgment is, that all lands which are worth plowing, which is not the case with all lands that are plowed, would *be improved* by draining; but I know that our farmers are neither able nor ready to drain to that extent, nor do I insist that it would pay while land is so cheap, and labor and tile so dear as at present. Ultimately, I believe, we shall tile-drain nearly all our level, or moderately sloping lands, that are worth cultivation."

[89]

Whether land would be *improved* by drainage, is one question, and whether the operation will *pay*, is quite another. The question whether it will pay, depends on the value of the land before drainage, the cost of the operation, and the value of the land when completed. And the cost of the operation includes always, not only the money and labor expended in it, but also the loss to other land of the owner, by diverting from it the capital which would otherwise be applied to it. Where labor and capital are limited so closely as they are in all our new States, it is a question not only how can they be profitably applied, but how can they be *most* profitably applied. A proprietor, who has money to loan at six per cent. interest, may well invest it in draining his land; when a working man, who is paying twelve per cent. interest for all the capital he employs, might ruin himself by making the same improvement.

## DO ALL LANDS REQUIRE DRAINAGE?

Our opinion is, that a great deal of land does not in any sense require drainage, and we should differ with Mr. Greeley, in the opinion that *all* lands worth ploughing, would be improved by drainage. Nature has herself thoroughly drained a large proportion of the soil. There is a great deal of finely-cultivated land in England, renting at from five to ten dollars per acre, that is thought there to require no drainage.

In a published table of estimates by Mr. Denton, made in 1855, it is supposed that Great Britain, including England, Scotland, and Wales, contain 43,958,000 acres of land, cultivated and capable of cultivation; of which he sets down as "wet land," or land requiring drainage, 22,890,004 acres, or about one half the whole quantity. His estimate is, that only about 1,365,000 acres had then been permanently drained, and that it would cost about [90] 107 millions of pounds to complete the operation, estimating the cost at about twenty shillings, or five dollars per acre.

These estimates are valuable in various views of our subject. They answer with some definiteness the question so often asked, whether all lands require drainage, and they tend to correct the impression, which is prevalent in this country, that there is something in the climate of Great Britain that makes drainage there essential to good cultivation on any land. The fact is not so. There, as in America, it depends upon the condition and character of the soil, more than upon the quantity of rain, or any condition of climate, whether drainage is required or not. Generally, it will be found on investigation, that so far as climate, including of course the quantity and regularity of the rain-fall, is concerned, drainage is more necessary in America than in Great Britain—the quantity of rain being in general greater in America, and far less regular in its fall. This subject, however, will receive a more careful consideration in another place.

If in America, as in Great Britain, one half the cultivable land require drainage, or even if but a tenth of that half require it, the subject is of vast importance, and it is no less important for us to apprehend clearly what part of our land does *not* require this expenditure, than to learn how to treat properly that which does require it.

To resume the inquiry, what lands require drainage? it may be answered—

## ALL LANDS OVERFLOWED IN SUMMER REQUIRE DRAINAGE.

Lands overflowed by the regular tides of the ocean require drainage, whether they lie upon the sea-shore, or upon rivers or bays. But this drainage involves embankments, and a peculiar mode of procedure, of which it is not now proposed to treat. [91]

Again, all lands overflowed by Summer freshets, as upon rivers and smaller streams, require drainage. These, too, usually require embankments, and excavations of channels or outlets, not within the usual scope of what is termed thorough drainage. For a further answer to the question—what lands require drainage? the reader is referred to the chapters which treat of the effect of drainage upon the soil.

## SWAMPS AND BOGS REQUIRE DRAINAGE.

No argument is necessary to convince rational men that the very extensive tracts of land, which are usually known as swamps and bogs, must, in some way, be relieved of their surplus water, before they can be rendered fit for cultivation. The treatment of this class of wet lands is so different from that applied to what we term upland, that it will be found more convenient to pass the subject by with this allusion, at present, and consider it more systematically under a separate head.

## ALL HIGH LANDS THAT CONTAIN TOO MUCH WATER AT ANY SEASON, REQUIRE DRAINAGE.

Draining has been defined, "The art of rendering land not only so free of moisture as that no superfluous water shall remain in it, but that no water shall remain in it so long as to injure, or even retard the healthy growth of plants required for the use of man and beast."

Some plants grow in water. Some even spring from the bottom of ponds, and have no other life than such a position affords. But most plants, useful to man, are drowned by being overflowed even for a short time, and are injured by any stagnant water about their roots. Why this is so, it is not easy to explain. Most of our knowledge on these points, is derived from observation. We know that fishes live in water, and if we would propagate [92] them, we prepare ponds and streams for the purpose. Our domestic animals live on land, and we do not put them into fish-ponds to pasture. There are useful plants which thrive best in water. Such is the cranberry, notwithstanding all that has been said of its cultivation on upland. And there are domestic fowls, such as ducks and geese, that require pools of water; but we do not hence infer that our hens and chickens would be better for daily immersion. All lands, then, require drainage, that contain too much water, at any season *for the intended crops.*

This will be found to be an important element in our rule. Land may require drainage for Indian corn, that may not require it for grass. Most of the cultivated grasses are improved in quality, and not lessened in quantity, by the removal of stagnant water in Summer; but there are reasons for drainage for hoed crops, which do not apply to our mowing fields. In New England, we have for a few weeks a perfect race with Nature, to get our seeds into the ground before it is too late. Drained land may be plowed and planted several weeks earlier than land undrained, and this additional time for preparation is of great value to the farmer. Much of this same land would be, by the first of June, by the time the ordinary planting season is past, sufficiently drained by Nature, and a grass crop upon it would be, perhaps, not at all benefitted by thorough-drainage; so that it is often an important consideration with reference to this operation, whether a given portion of our farm may not be most profitably kept in permanent grass, and maintained in fertility by top-dressing, or by occasional plowing and reseeding in Autumn. It is certainly convenient to have all our fields adapted to our usual rotation, and it is for each man to balance for himself this convenience against the cost of drainage in each particular case.

What particular crops are most injured by stagnant [93] water in the soil, or by the too tardy percolation of rain-water, may be determined by observation. How stagnant water injures plants, is not, as has been suggested, easily understood in all its relations. It doubtless retards the decomposition of the substances which supply their nutriment, and it reduces the temperature of the soil. It has been suggested, that it prevents or checks perspiration and introsusception, and it excludes the air which is essential to the vegetation of most plants. Whatever the theory, the fact is acknowledged, that stagnant water *in* as well as *on* the soil, impedes the growth of all our valuable crops, and that drainage soon cures the evil, by removing the effect with its cause. And the remedy seems to be almost instantaneous; for, on most upland, it is found that by the removal of stagnant water, the soil is in a single season rendered fit for the growth of cultivated crops. In low meadows, composed of peat and swamp mud, in many cases, exposure to the air for a year or two after drainage, is often found to enhance the fertility of the soil, which contains, frequently, acids which need correction.

## INDICATIONS OF TOO MUCH MOISTURE.

It has already been suggested, that motives of convenience may induce us to drain our lands—that we may have a longer season in which to work them; and that there may be cases where the crop would flourish if planted at precisely the right time, where yet we cannot well, without drainage, seasonably prepare for the crop. Generally, however, lands too wet seasonably to plant, will give indications, throughout the season, of hidden water producing its ill effects.

If the land be in grass, we find that aquatic plants, like rushes or water grasses, spring up with the seeds we have sown, and, in a few years, have possession of the field, and we are soon compelled to plow up the sod, and lay [94] it again to grass. If it be in wheat or other grain, we see the field spotted and uneven; here a portion on some slight elevation, tall and dark colored, and healthy; and there a little depression, sparsely covered with a low and sickly growth. An American traveling in England in the growing season, will always be struck with the perfect *evenness* of the fields of grain upon the well-drained soil. Journeying through a considerable portion of England and Wales with intelligent English farmers, we were struck with their nice perception on this point.

The slightest variation in the color of the wheat in the same or different fields, attracted their instant attention.

"That field is not well-drained; the corn is too light-colored." "There is cold water at the bottom there; the corn cannot grow;" were the constant criticisms, as we passed across the country. Inequalities that, in our more careless cultivation, we should pass by without observation, were at once explained by reference to the condition of the land, as to water.

The drill-sowing of wheat, and the careful weeding it with the horse-hoe and by hand, are additional reasons why the English fields should present a uniform appearance, and why any inequalities should be fairly referable to the condition of the soil.

Upon a crop of Indian corn, the cold water lurking below soon places its unmistakable mark. The blade comes up yellow and feeble. It takes courage in a few days of bright sunshine in June, and tries to look hopeful, but a shower or an east wind again checks it. It had already more trouble than it could bear, and turns pale again. Tropical July and August induce it to throw up a feeble stalk, and to attempt to spindle and silk, like other corn. It goes through all the forms of vegetation, and yields at last a single nubbin for the pig. Indian corn [95] must have land that is dry in Summer, or it cannot repay the labor of cultivation.

Careful attention to the subject will soon teach any farmer what parts of his land are injured by too much water; and having determined that, the next question should be, whether the improvement of it by drainage will justify the cost of the operation.

## WILL IT PAY?

Drainage is a permanent investment. It is not an operation like the application of manure, which we should expect to see returned in the form of salable crops in one or two years, or ten at most, nor like the labor applied in cultivating an annual crop. The question is not whether drainage will pay in one or two years, but will it pay in the long run? Will it, when completed, return to the farmer a fair rate of interest for the money expended? Will it be more profitable, on the whole, than an investment in bank or railway shares, or the purchase of Western lands? Or, to put the question in the form in which an English land-owner would put it, will the rent of the land improved by drainage, be permanently increased enough to pay a fair interest on the cost of the improvement?

Let us bring out this idea clearly to the American farmer by a familiar illustration. Your field is worth to you now one hundred dollars an acre. It pays you, in a series of years, through a rotation of planting, sowing, and grass, a nett profit of six dollars an acre, above all expenses of cultivation and care.

Suppose, now, it will cost one-third of a hundred dollars an acre to drain it, and you expend on each three acres one hundred dollars, what must the increase of your crops be, to make this a fair investment? Had you expended the hundred dollars in *labor*, to produce a crop of [96] cabbages, you ought to get your money all back, with a fair profit, the first year. Had you expended it in guano or other special manures, whose beneficial properties are exhausted in some two or three years, your expenditure should be returned within that period. But the improvement by drainage is permanent; it is done for all time to come. If, therefore, your drained land shall pay you a fair rate of interest on the cost of drainage, it is a good investment. Six per cent. is the most common rate of interest, and if, therefore, each three acres of your drained land shall pay you an increased annual income of six dollars, your money is fairly invested. This is at the rate of two dollars an acre. How much increase of crop will pay this two dollars? In the common rotation of Indian corn, potatoes, oats, wheat, or barley, and grass, two or three bushels of corn, five or six bushels of potatoes, as many bushels of oats, a bushel or two of wheat, two or three bushels of barley, will pay the two dollars. Who, that has been kept back in his Spring's work by the wetness of his land, or has been compelled to re-plant because his seed has rotted in the ground, or has experienced any of the troubles incident to cold wet seasons, will not admit at once, that any land which Nature has not herself thoroughly drained, will, in this view, pay for such improvement?

But far more than this is claimed for drainage. In England, where such operations have been reduced to a system, careful estimates have been made, not only of the cost of drainage, but of the increase of crops by reason of the operation.

In answer to questions proposed by a Board of Commissioners, in 1848, to persons of the highest reputation for knowledge on this point, the increase of crops by drainage was variously stated, but in no case at less than a paying rate. One gentleman says: "A sixth of increase in [97] produce of grain crops may be taken as the very lowest estimate, and, in actual result, it is seldom less than one-fourth. In very many cases, after some following cultivation, the produce is doubled, whilst the expense of working the land is much lessened." Another says: "In many instances, a return of fully 25 per cent. on the expenditure is realized, and in some even more." A third remarks, "My experience and observation have chiefly been in heavy clay soils, where the result of drainage is eminently beneficial, and where I should estimate the increased crop at six to ten bushels (wheat) per statute acre."

These are estimates made upon lands that had already been under cultivation. In addition to such lands as are merely rendered less productive by surplus water, we have, even on our hard New England farms — on side hills, where springs burst out, or at the foot of declivities, where the land is flat, or in runs, which receive the natural drainage of higher lands — many places which are absolutely unfit for cultivation, and worse than useless, because they separate those parts of the farm which can be cultivated. If, of these wet portions, we make by draining, good, warm, arable land, it is not a mere question of per centage or profit; it is simply the question whether the land, when drained, is worth more than the cost of drainage. If it be, how much more satisfactory, and how much more profitable it is, to expend money in thus reclaiming the waste places of our farms, and so uniting the detached fields into a compact, systematic whole, than to follow the natural bent of American minds, and "annex" our neighbor's fields by purchasing.

Any number of instances could be given of the increased value of lands in England by drainage, but they are of little practical value. The facts, that the Government has made large loans in aid of the process, that private [98] drainage companies are executing extensive works all over the kingdom, and that large land-holders are draining at their own cost, are conclusive evidence to any rational mind, that drainage in Great Britain, at least, well repays the cost of the operation.

In another chapter may be found accurate statements of American farmers of their drainage operations, in different States, from which the reader will be able to form a correct opinion, whether draining in this country is likely to prove a profitable operation.

# CHAPTER V [99]
# VARIOUS METHODS OF DRAINAGE.

Open Ditches.—Slope of Banks.—Brush Drains.—Ridge and Furrow.—Plug-Draining.—Mole-Draining.—Mole-Plow.—Wedge and Shoulder Drains.—Larch Tubes.—Drains of Fence Rails, and Poles.—Peat Tiles.—Stone Drains Injured by Moles.—Downing's Giraffes.—Illustrations of Various Kinds of Stone Drains.

## OPEN DITCHES.

The most obvious mode of getting rid of surface-water is, to cut a ditch on the surface to a lower place, and let it run. So, if the only object were to drain a piece of land merely for a temporary purpose—as, where land is too wet to ditch properly in the first instance, and it is necessary to draw off part of the surplus water before systematic operations are commenced—an open ditch is, perhaps, the cheapest method to be adopted.

Again: where land to be drained is part of a large sloping tract, and water runs down, at certain seasons, in large quantities upon the surface, an open catch-water-ditch may be absolutely necessary. This condition of circumstances is very common in mountainous districts, where the rain which falls on the hills flows down, either on the visible surface or on the rock-formation under the soil, and breaks out at the foot, causing swamps, often high up on the hillsides. Often, too, in clay districts, where sand or loam two or three feet deep rests on tough clay, we see broad sloping tracts, which form our best grass-fields.

If we are attempting to drain the lower part of such a [100] slope, we shall find that the water from the upper part flows down in large quantities upon us, and an open ditch may be most economical as a header, to cut off the down-flowing water; though, in most cases, a covered drain may be as efficient.

At the outlets, too, of our tile or stone drains, when we come down nearly to the level of the stream which receives our drainage-water, we find it convenient, often, and indeed necessary, to use open ditches—perhaps only a foot or two deep—to carry off the water discharged. These ditches are of great importance, and should be finished with care, because, if they become obstructed, they cause back-water in the drains, and may ruin the whole work.

Open drains are thus essential auxiliaries to the best plans of thorough drainage; and, whatever opinion may be entertained of their economy, many farmers are so situated that they feel obliged to resort to them for the present, or abandon all idea of draining their wet lands. We will, therefore, give some hints as to the best manner of constructing open drains; and then suggest, in the form of objections to them, such considerations as shall lead the proprietor who adopts this mode to consider carefully his plan of operations in the outset, with a view to obviate, as much as possible, the manifest embarrassments occasioned by them.

As to the location of drains in swamps and peculiarly wet places, directions may be found in another chapter. We here propose only to treat of the mode of forming open drains, after their location is fixed.

The worst of all drains is an open ditch, of equal width from top to bottom. It cannot stand a single season, in any climate or soil, without being seriously impaired by the frosts or the heavy rains. All open drains should be sloping; and it is ascertained, by experiment, what is the [101] best, or, as it is sometimes expressed, the natural slope, on different kinds of soil. If earth be tipped from a cart down a bank, and be left exposed to the action of the weather, it will rest, and finally remain, at a regular angle or inclination, varying from 21° to 55° with the horizon, according to the nature of the soil. The natural slope of common earth is found to be about 33° 42'; and this is the inclination usually adopted by railroad engineers for their embankments.

If the banks of the open ditch are thus sloped, they will have the least possible tendency to wash away, or break down by frost.

Again: where open ditches are adopted in mowing fields, they may, if not very deep, be sloped still lower than the natural slope, and seeded down to the bottom; so that no land will be lost, and so that teams may pass across them.

This amounts, in fact, to the old ridge and furrow system, which was almost universal in England before tiles were used, and is sometimes seen practiced in this country. The land, by that system, is back-furrowed in narrow lands, till it is laid up into beds, sloping from the tops, or backs, to the furrows which constitute the drains. This mode of culture is very ancient, and is probably referred to in the language of the Psalmist, in the Scriptures: "Thou waterest the ridges thereof abundantly, thou settlest the furrows thereof, thou makest it soft with showers."

The objections to open ditches, as compared with under-drains, may be briefly stated thus:

1. *They are expensive.* The excavation of a sloping drain is much greater than that of an upright drain. An open drain must have a width of one or two feet at the bottom, to receive the earth that always must, to some extent, wash into it. An open drain requires to be cleaned out once a year, to keep it in good order. There is a large [102] quantity of earth from an open drain to be disposed of, either by spreading or hauling away. Thus, a drain of this kind is costly at the outset, and requires constant labor and care to preserve it in working condition.

2. *They are not permanent.* A properly laid underdrain will last half a century or more, but an open drain, especially if deep, has a constant tendency to fill up. Besides, the action of frost and water and vegetation has a continual operation to obstruct open ditches. Rushes and water-grasses spring up luxuriantly in the wet and slimy bottom, and often, in a single season, retard the flow of water, so that it will stand many inches deep where the fall is slight. The slightest accident, as the treading of cattle, the track of a loaded cart, the burrowing of animals, dams up the water and lessens the effect of the drain. Hence, we so often see meadows which have been drained in this way going back, in a few years, into wild grass and rushes.

3. *They obstruct good husbandry.* In the chapter upon the effects of drainage on the condition of the soil, we suggest, in detail, the hindrances which open ditches present to the convenient cultivation of the land, and, especially, how they obstruct the farmer in his plow-

ing, his mowing, his raking, and the general laying out of his land for convenient culture.

4. *They occupy too much land.* If a ditch have an upright bank, it is so soft that cattle will not step within several feet of it in plowing, and thus a strip is lost for culture, or must be broken up by hand. If, indeed, we can get the plow near it, there being no land to rest against, the last furrow cannot be turned from the ditch, and if it be turned into it, must be thrown out by hand. If the banks be sloped to the bottom, and the land be thus laid into beds or ridges, the appearance of the field may, indeed, be improved, but there is still a loss of soil; [103] for the soil is all removed from the furrow, which will always produce rushes and water-grass, and carried to the ridge, where it doubles the depth of the natural soil. Thus, instead of a field of uniform condition, as to moisture and temperature and fertility, we have strips of wet, cold, and poor soil, alternating with dry, warm, and rich soil, establishing a sort of gridiron system, neither beautiful, convenient, nor profitable.

5. *The manure washes off and is lost.* The three or four feet of water which the clouds annually give us in rain and snow, must either go off by evaporation, or by filtration, or run off upon the surface. Under the title of Rain and Evaporation, it will be seen that not much more than half this quantity goes off by evaporation, leaving a vast quantity to pass off through or upon the soil. If lands are ridged up, the manure and finer portions of the soil are, to a great extent, washed away into the open ditches and lost. Of the water which filters downwards, a large portion enters open ditches near the surface, before the fertilizing elements have been strained out; whereas, in covered drains of proper depth, the water is filtered through a mass of soil sufficiently deep to take from it the fertilizing substances, and discharge it, comparatively pure, from the field. In a paper by Prof. Way (11th Jour. Roy. Ag. Soc.), on "The Power of Soils to retain Manure," will be found interesting illustrations of the filtering qualities of different kinds of soil.

In addition to the above reasons for preferring covered drains, it has been asserted by one of the most skillful drainers in the world (Mr. Parkes), "that a proper covered drain of the same depth as an open ditch, will drain a greater breadth of land than the ditch can effect. The sides of the ditch," he says, "become dried and plastered, and covered with vegetation; and even while they are [104] free from vegetation, their absorptive power is inferior to the covered drain."

Of the depth, direction, and distance of drains, our views will be found under the appropriate heads. They apply alike to open and covered drains.

## BRUSH DRAINS.

Having a farm destitute of stones, before tiles were known among us, we made several experiments with covered drains filled with brush. Some of those drains operated well for eight or ten years; others caved in and became useless in three or four years, according to the condition of the soil.

In a wet swamp a brush drain endures much longer than in sandy land, which is dry a part of the year, because the brush decays in dry land, but will prove nearly imperishable in land constantly wet. In a peat or muck swamp, we should expect that such drains, if carefully constructed, might last twenty years, but that in a sandy loam they would be quite unreliable for a single year.

Our failure on upland with brush drains, has resulted, not from the decay of the wood, but from the entrance of sand, which obstructed the channel. Moles and field-mice find these drains the very day they are laid, and occupy them as permanent homes ever after.

Those little animals live partly upon earth-worms, which they find by burrowing after them in the ground, and partly upon insects, and vegetation above ground. They have a great deal of business, which requires convenient passages leading from their burrows to the day-light, and drains in which they live will always be found perforated with holes from the surface. In the Spring, or in heavy showers, the water runs in streams into these holes, breaks down the soft soil as it goes, and finally the top begins to fall in, and the channel is choked up, and the work ruined. [105] We have tried many precautions against this kind of accident, but none that was effectual on light land.

The general mode of construction is this: Open the trench to the depth required, and about 12 inches wide at the bottom. Lay into this poles of four or five inches diameter at the butt, leaving an open passage between. Then lay in brush of any size, the coarsest at the bottom, filling the drain to within a foot of the surface, and covering with pine, or hemlock, or spruce boughs. Upon these lay turf, carefully cut, as close as possible. The brush should be laid but-end up stream, as it obstructs the water less in this way. Fill up with soil a foot above the surface, and tread it in as hard as possible. The weight of earth will compress the brush, and the surface will settle very much. We have tried placing boards at the sides, and upon the top of the brash, to prevent the caving in, but with no great success. Although our drains thus laid, have generally continued to discharge some water, yet they have, upon upland, been dangerous traps and pitfalls for our horses and cattle, and have cost much labor to fill up the holes, where they have fallen through by washing away below.

In clay, brush drains might be more durable. In the English books, we have descriptions of drains filled with thorn cuttings from hedges and with gorse. When well laid in clay, they are said to last about 15 years. When the thorns decay, the clay will still retain its form, and leave a passage for the water.

A writer in the Cyclopedia sums up the matter as to this kind of drains, thus:

"Although in some districts they are still employed, they can only be looked upon as a clumsy, and superficial plan of doing that which can be executed in a permanent and satisfactory manner, at a very small additional expense, now that draining-tiles are so cheap and plentiful."

Draining-tiles are not yet either cheap or plentiful in [106] this country; but we have full faith that they will become so very soon. In the mean time it may be profitable for us to use such of the substitutes for them as may lie within our reach, selecting one or another according as material is convenient.

## PLUG-DRAINING

has never been, that we are aware, practiced in America. Our knowledge of it is limited to what we learn from English books. We, therefore, content ourselves with giving from Morton's Cyclopedia the following description and illustrations.

"*Plug-draining*, like mole-draining, does not require the use of any foreign material—the channel for the water being wholly formed of clay, to which this kind of drain, like that last mentioned is alone suited.

"This method of draining requires a particular set of tools for its execution, consisting of, first, a common spade, by means of which the first spit is removed, and laid on one side; second, a smaller-sized spade, by means of which the second spit is taken out, and laid on the opposite side of the trench thus formed; third, a peculiar instrument called a bitting iron (Fig. 11), consisting of a narrow spade, three and a half feet in length, and one and a half inches wide at the mouth and sharpened like a chisel; the mouth, or blade, being half an inch in thickness in order to give the necessary strength to so slender an implement. From the mouth, *a*, on the right-hand side, a ring of steel, *b*, six inches long and two and a half broad, projects at right angles; and on the left, at fourteen inches from the mouth, a tread, *c*, three inches long, is fitted.

Fig. 11.

"A number of blocks of wood, each one foot long, six inches high, and two inches thick at the bottom, and two and a half at the top, are next required. From four to six of these are joined together by pieces [107] of hoop-iron let into their sides by a saw-draught, a small space being left between their ends, so that when completed, the whole forms a somewhat flexible bar, as shown in the cut, to one end of which a stout chain is attached. These blocks are wetted, and placed with the narrow end undermost, in the bottom of the trench, which should be cut so as to fit them closely; the clay which has been dug out is then to be returned, by degrees, upon the blocks, and rammed down with a wooden rammer three inches wide. As soon as the portion of the trench above the blocks, or plugs, has been filled, they are drawn forward, by means of a lever thrust through a link of the chain, and into the bottom of the drain for a fulcrum, until they are all again exposed, except the last one. The further portion of the trench, above the blocks, is now filled in and rammed, and so on the operations proceed until the whole drain is finished."

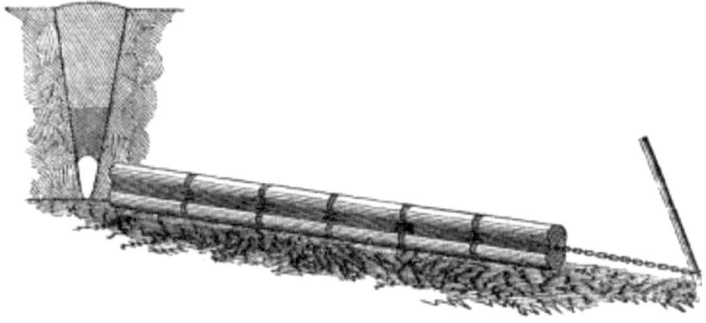

Fig. 12.—Plug Drainage.

## MOLE DRAINING.

We hear of an implement, in use in Illinois and other Western States, called the Gopher Plow, worked by a capstan, which drains wet land by merely drawing through it an iron shoe, at about two and a half feet in depth, without the use of any foreign substance.

We hear reports of a mole plow, in use in the same State, known by the name of Marcus and Emerson's Patent Subsoiler, with which, an informant says, drains are made also in the manner above named. This machine [108] is worked by a windlass power, by a horse or yoke of oxen, and the price charged is twenty-eight cents a rod for the work. These machines are, from description, modifications of the English Mole Plow, an implement long ago known and used in Great Britain.

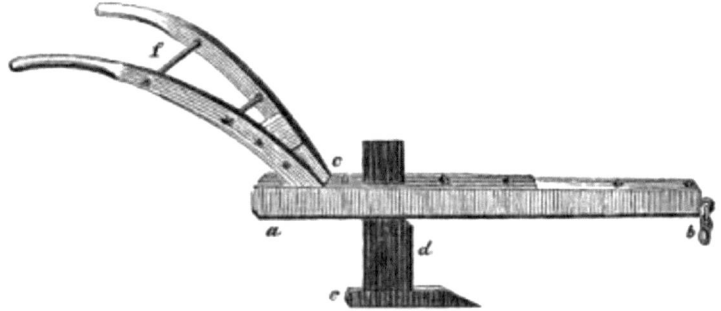

Fig. 13.—Mole Plow.

The following description is from Morton's Cyclopedia:

"*Mole-Drains* are the simplest of all the forms of the covered drains. They are formed by means of a machine called the mole plow. This machine consists of a long wooden beam and stilts, somewhat in the form of the subsoil plow; but instead of the apparatus for breaking up the subsoil in the latter, a short cylindrical and pointed bar of iron is attached, horizontally, to the lower end of the broad coulter, which can be raised or lowered by means of a slot in the beam. The beam itself is sheathed with iron on the under side, and moves close to the ground; thus keeping the bar at the end of the coulter at one uniform depth. This machine is dragged through the soft clay, which is the only kind of land on which it can be used with propriety, by means of a chain and capstan, worked by horses, and produces a hollow channel very similar to a mole-run, from which it derives its name."

A correspondent of the *New York Tribune* thus describes the operation and utility of a mole plow, which he saw on the farm of Major A. B. Dickinson, of Hornby, Steuben County, New York:

"I believe there is not a rod of tile laid on this farm, and not a dozen [109] rods of covered stone drain. But the major has a home-made, or, at least, home-devised, 'bull plow,' consisting of a sharp-pointed iron wedge, or roller, surmounted by a broad, sharp shank nearly four feet high, with a still sharper cutter in front, and with a beam and handles above all. With five yoke of oxen attached, this plow is put down through the soil and subsoil to an average depth of three feet—in the course which the superfluous water is expected and desired to take—and the field thus plowed through and through, at intervals of two rods, down to three feet, as the ground is more or less springy and saturated with water. The cut made by the shank closes after the plow and is soon obliterated, while that made by the roller, or wedge, at the bottom, becomes the channel of a stream of water whenever there is any excess of moisture above its level, which stream tends to clear itself and rather enlarge its channel. From ten to twenty acres a day are thus drained, and Major D. has such drains of fifteen to twenty years' standing, which still do good service. In rocky soils, this mode of draining is impracticable: in sandy tracts it would not endure; but here it does very well, and, even though it should hold good in the average but ten years, it would many times repay its cost."

Major Dickinson himself in a recent address, thus speaks of what he calls his

## SHANGHAE PLOW.

"I will take the poorest acre of stubble ground, and if too wet for corn in the first place, I will thoroughly drain it with a Shanghae plow and four yoke of oxen in three hours.

"I will suppose the acre to be twenty rods long and eight rods wide. To thoroughly drain the worst of your clay subsoil, it may require a drain once in eight feet, and they can be made so cheaply that I can afford to make them at that distance. To do so, will require the team to travel sixteen times over the twenty rods lengthwise, or one mile in three hours; two men to drive, one to hold the plow, one to ride the beam, and one to carry the crow-bar, pick up any large stones thrown out by going to the right or left, and to help to carry around the plow, which is too heavy for the other two to do quickly.

"The plow is quite simple in its construction, consisting of a round piece of iron three and a half or four inches in diameter, drawn down to a point, with a furrow cut in the top one and a half inches deep; a plate, eighteen inches wide and three feet long, with one end welded into the furrow of the round bar, while the other is fastened to the [110] beam. The coulter is six inches in width, and is fastened to the beam at one end, and at the other to the point of the round bar. The coulter and plate are each three-fourths of an inch thick, which is the entire width of the plow above the round iron at the bottom.

"It would require much more team to draw this plow on some soils than on yours. The strength of team depends entirely on the character of the subsoil. Cast-iron, with the exception of the coulter, for an easy soil would be equally good; and from eighteen to twenty-four inches is sufficiently deep to run the plow. I can as thoroughly drain an acre of ground in this way as any that can be found in Seneca County."

From the best information we can gather, it would seem, that on certain soils with a clay subsoil, the mole plow, as a sort of pioneer implement, may be very useful. The above account certainly indicates that on the farm in question it is very cheap, rapid, and effectual in its operation.

Stephens gives a minute description of the mole plow figured above, in his Book of the Farm. Its general structure and principle of operation may be easily understood by what has been already said, and any person desirous of constructing one may find in that work exact directions.

## WEDGE AND SHOULDER DRAINS.

These, like the last-mentioned kind of drains, are mere channels formed in the subsoil. They have, therefore, the same fault of want of durability, and are totally unfitted for land under the plow. In forming *wedge-drains*, the first spit, with the turf attached, is laid on one side, and the earth removed from the remainder of the trench is laid on the other. The last spade used is very narrow, and tapers rapidly, so as to form a narrow wedge-shaped cavity for the bottom of the trench. The turf first removed is then cut into a wedge, so much larger than the size of the lower part of the drain, that when rammed into it with the grassy side undermost, it leaves a vacant space in the bottom six or eight inches in depth, as in Fig. 14.

The *shoulder-drain* does not differ very materially from [111] the wedge-drain. Instead of the whole trench forming a gradually tapering wedge, the upper portion of the shoulder-drain has the sides of the trench nearly perpendicular, and of considerable width, the last spit only being taken out with a narrow, tapering spade, by which means a shoulder is left on either side, from which it takes its name. After the trench has been finished, the first spit, having the grassy side undermost as in the former case, is placed in the trench, and pushed down till it rests upon the shoulders already mentioned; so that a narrow wedge-shaped channel is again left for the water, as shown in Fig. 15.

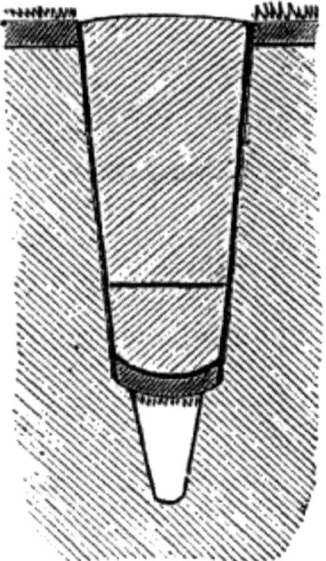

Fig. 14.
Wedge-Drain.

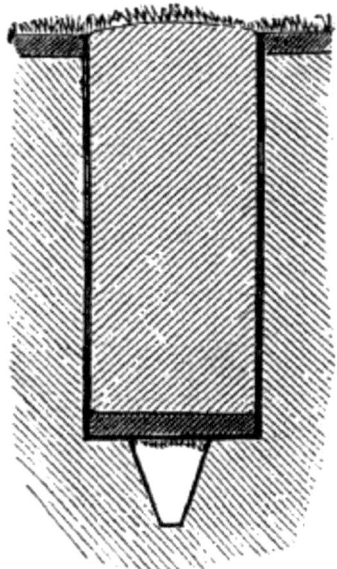

Fig. 15.
Shoulder-Drain.

These drains may be formed in almost any kind of land which is not a loose gravel or sand. They are a very cheap kind of drain; for neither the cost of cutting nor filling in, much exceeds that of the ordinary tile drain, while the expense of tiles or other materials is altogether saved. Still, such drains cannot be recommended, for they are very liable to injury, and, even under the most favorable circumstances, can only last a very limited time.

## LARCH TUBES.

These have been used in Scotland, in mossy or swampy soils, it is said, with economy and good results. The tube [112] represented below presents a square of 4 inches outside, with a clear water-way of 2 inches. Any other durable wood will, of course, answer the same purpose. The tube is pierced with holes to admit the water. In wet meadows, these tubes laid deep would be durable and efficient, and far more reliable than brush or even stones, because they may be better protected from the admission of sand and the ruinous working of vermin. Their economy depends upon the price of the wood and the cost of tiles—which are far better if they can be reasonably obtained.

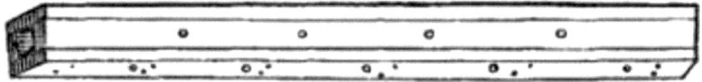

Fig. 16.—Larch Tube-Drain.

Near Washington, D. C., we know of drainage tolerably well performed by the use of common fence-rails. A trench is opened about three inches wider at bottom than two rails. Two rails are then laid in the bottom, leaving a space of two or three inches between them. A third rail is then laid on for a cover, and the whole carefully covered with turf or straw, and then filled up with earth. Poles of any kind may be used instead of rails, if more convenient.

In clay, these drains would be efficient and durable; in sand, they would be likely to be filled up and become useless. This is an extravagant waste of timber, except in the new districts where it is of no value.

Mr. J. F. Anderson, of Windham, Maine, has adopted a mode of draining with poles, which, in regions where wood is cheap and tiles are dear, may be adopted with advantage.

Two poles, of from 3 to 6 inches diameter, are laid at the bottom of the ditch, with a water-way of half their diameter between them. Upon these, a third pole is laid, [113] thus forming a duct of the desired dimensions. The security of this drain will depend upon the care with which it is protected by a covering of turf and the like, to prevent the admission of earth, and its permanency will depend much upon its being placed low enough to be constantly wet, as such materials are short-lived when frequently wet and dried, and nearly imperishable if constantly wet. It is unnecessary to place brush or stones over such drains to make them draw, as it is called. The water will find admission fast enough to destroy the work, unless great care is used.

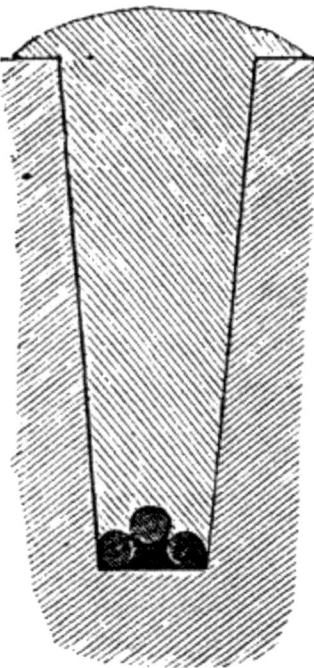

Fig. 17.—Pole-Drain.

In Ireland, and in some parts of England and Scotland, peat-tiles are sometimes used in draining bogs. They are cheap and very durable in such localities, but, probably, will not be used in this country. They are formed somewhat like pipes, of two pieces of peat. Two halves are formed with a peculiar tool, with a half circle in each. When well dried, they are placed together, thus making a round opening.

Fig. 18.—Tool for Peat-Tiles.

Fig. 19.—Peat-Tiles.

In draining, the object being merely to form a durable [114] opening in the soil, at suitable depth, which will receive and conduct away the water which filters through the soil, it is obvious that a thousand expedients may be resorted to, to suit the peculiar circumstances of persons. In general, the danger to be apprehended is from obstruction of the water-way. Nothing, except a tight tube of metal or wood, will be likely to prevent the admission of water.

Economy and durability are, perhaps, the main considerations. Tiles, at fair prices, combine these qualities better than anything else. Stones, however, are both cheap and durable, so far as the material is concerned; but the durability of the material, and the durability of the drains, are quite different matters.

## DRAINS OF STONES.

Providence has so liberally supplied the greater part of New England with stones, that it seems to most inexperienced persons to be a work of supererogation, almost, to manufacture tiles or any other draining material for our farms.

We would by no means discourage the use of stones, where tiles cannot be used with greater economy. Stone drains are, doubtless, as efficient as any, so long as the water-way can be kept open. The material is often close at hand, lying on the field and to be removed as a nuisance, if not used in drainage. In such cases, true economy may dictate the use of them, even where tiles can be procured; though, we believe, tiles will be found generally cheaper, all things considered, where made in the neighborhood.

In treating of the cost of drainage, we have undertaken to give fair estimates of the comparative cost of different materials.

Every farmer is capable of making estimates for himself, [115] and of testing those made by us, and so of determining what is true economy in his particular case.

The various modes of constructing drains of stones, may be readily shown by simple illustrations:

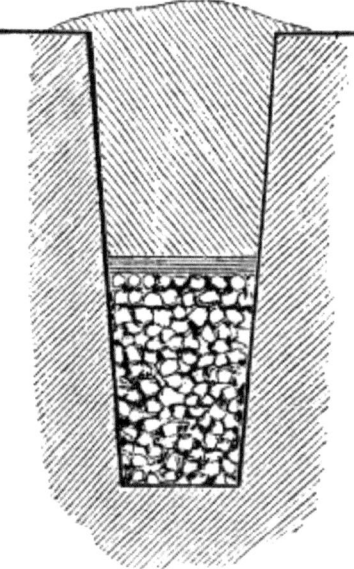

Fig. 20.

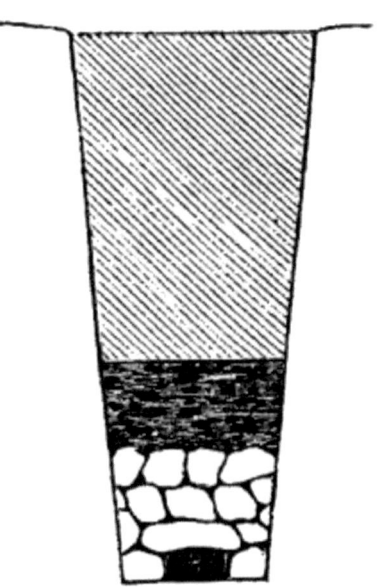

Fig. 21.

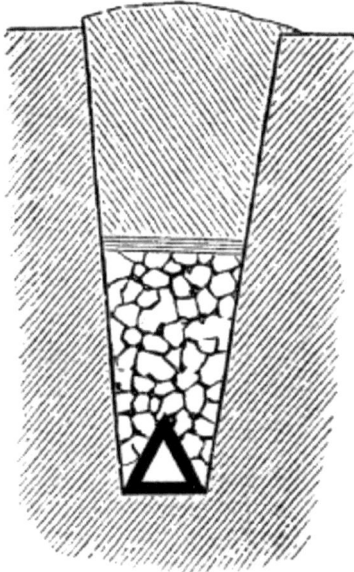

Fig. 22.

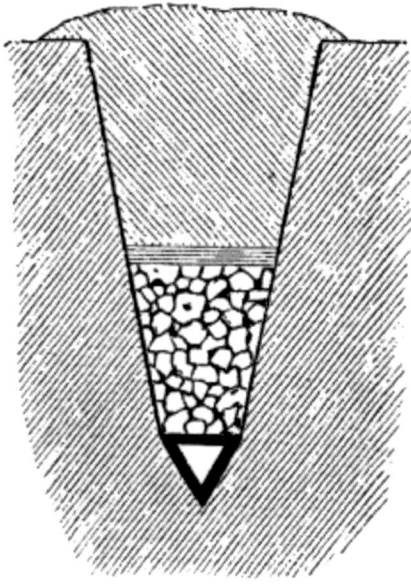

Fig. 23.

If stone-drains are decided upon, the mode of constructing them will depend upon the kind of stone at hand. In some localities, round pebble-stones are found scattered over the surface, or piled in heaps upon our farms; in others, flat, slaty stones abound, and in others, broken stones from quarries may be more convenient. Of these, probably, [116] the least reliable is the drain filled with pebble-stones, or broken stones of small size. They are peculiarly liable to be obstructed, because there is no regular water-way, and the flow of the water must, of course, be very slow, impeded as it is by friction at all points with the irregular surfaces.

Sand, and other obstructing substances, which find their way, more or less, into all drains, are deposited among the stones — the water having no force of current sufficient to carry them forward — and the drain is soon filled up at some point, and ruined.

Miles of such drains have been laid on many New England farms, at shoal depths, of two or two and a half feet, and have in a few years failed. For a time, their effect, to those unaccustomed to under-drainage, seems almost miraculous. The wet field becomes dry, the wild grass gives place to clover and herds-grass, and the experiment is pronounced successful. After a few years, however, the wild grass re-appears, the water again stands on the surface, and it is ascertained, on examination, that the drain is in some place packed solid with earth, and is filled with stagnant water.

The fault is by no means wholly in the material. In clay or hard pan, such a drain may be made durable, with proper care, but it must be laid deep enough to be beyond the effect of the treading of cattle and of loaded teams, and the common action of frost. They can hardly be laid low enough to be beyond the reach of our great enemy, the mole, which follows relentlessly all our operations.

We recollect the remarks of Mr. Downing about the complaints in New England, of injury to fruit-trees by the gnawing of field-mice.

He said he should as soon think of danger from injury by giraffes as field-mice, in his own neighborhood, though he had no doubt of their depredations elsewhere!

It may seem to many, that we lay too much stress on [117] this point, of danger from moles and mice. We know whereof we do testify in this matter. We verily believe that we never finished a drain of brush or stones, on our farm, ten rods long, that there was not a colony of these *varmint* in the one end of it, before we had finished the other. If these drains, however, are made three or four feet deep, and the solid earth rammed hard over the turf, which covers the stones, they will be comparatively safe.

The figures 24 and 25 below, represent a mode of laying stone drains, practiced in Ireland, which will be found probably more convenient and secure than any other method, for common small drains. A flat stone is set upright against one side of the ditch, which should be near the bottom, perpendicular. Another stone is set leaning against the first, with its foot resting against the opposite bank. If the soil be soft clay, a flat stone may be placed first on the bottom of the ditch, for the water to flow upon; but this will be found a great addition to the labor, unless flat stones of peculiarly uniform shape and thickness are at hand. A board laid at the bottom will be usually far cheaper, and less liable to cause obstructions.

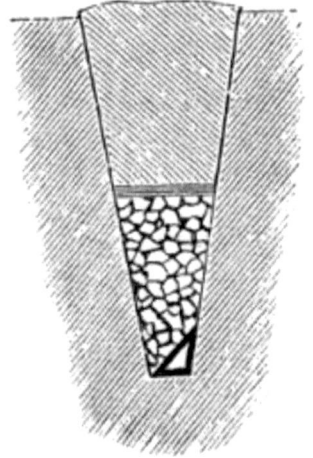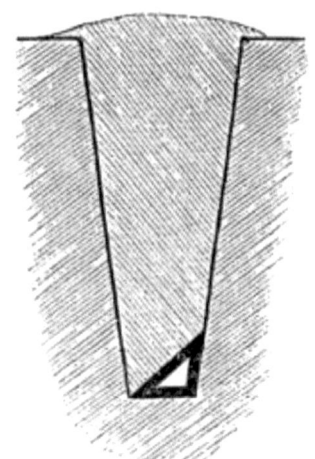

Figs. 24, 25.—Stone Drains.

Figure 25 represents the ditch without the small stones [118] above the duct. These small stones are, in nine cases in ten, worse than useless, for they are not only unnecessary to admit the water, but furnish a harbor for mice and other vermin.

Drawings, representing a filling of small stones above the duct, have been copied from one work to another for generations, and it seems never to have occurred, even to modern writers, that the small stones might be omitted. Any one, who knows anything of the present system of draining with tiles, must perceive at once that, if we have the open triangular duct or the square culvert, the water cannot be kept from finding it, by any filling over it with such earth as is usually found in ditching. Formerly, when tiles were used, the ditch was filled above the tiles, to the height of a foot or more, with broken stones; but this practice has been everywhere abandoned as expensive and useless.

An opening of any form, equal to a circle of two or three inches diameter, will be sufficient in most cases, though the necessary size of the duct must, of course, depend on the quantity of water which may be expected to flow in it at the time of the greatest flood.

Whatever the form of the stone drain, care should be taken to make the joints as close as possible, and turf, shavings, straw, tan, or some other material, should be carefully placed over the joints, to prevent the washing in of sand, which is the worst enemy of all drains.

It is not deemed necessary to remark particularly upon the mode of laying large drains for water-courses, with abutments and covering stones, forming a square duct, because it is the mode universally known and practiced. For small drains, in thorough-draining lands, it may, however, be remarked, that this is, perhaps, the most expensive of all modes, because a much greater width of excavation is necessary in order to place in position the two [119] side stones and leave the requisite space between them. That mode of drainage which requires the least excavation and the least carriage of materials, and consequently the least filling up and levelling, is usually the cheapest.

Our conclusion as to stone drains is, that, at present, they may be, in many cases, found useful and economical; and even where tiles are to be procured at present prices stones may well be used, where materials are at hand, for the largest drains.

# CHAPTER VI [120]
# DRAINAGE WITH TILES.

What are Drain-Tiles?—Forms of Tiles.—Pipes.—Horse-shoe Tiles.—Sole-Tiles—Form of Water-Passage.—Collars and their Use.—Size of Pipes.—Velocity.—Friction.—Discharge of Water through Pipes.—Tables of Capacity.—How Water enters Tiles.—Deep Drains run soonest and longest.—Pressure of Water on Pipes.—Durability of Tile Drains.—Drain-Bricks 100 years old.

## WHAT ARE DRAIN-TILES?

This would be an absurd question to place at the head of a division in a work intended for the English public, for tiles are as common in England as bricks, and their forms and uses as familiar to all. But in America, though tiles are used to a considerable extent in some localities, probably not one farmer in one hundred in the whole country ever saw one.

The author has recently received letters of inquiry about the use and cost of tiles, from which it is manifest that the writers have in their mind as tiles, the square bricks with which our grandfathers used to lay their hearths.

In Johnstone's *Report to the Board of Agriculture on Elkington's System of Draining*, published in England in 1797, the only kind of tiles or clay conduits described or alluded to by him, are what he calls "draining-bricks," of which he gives drawings, which we transfer to our pages precisely as found in the American edition. It will be [121] seen to be as clumsy a contrivance as could well be devised.

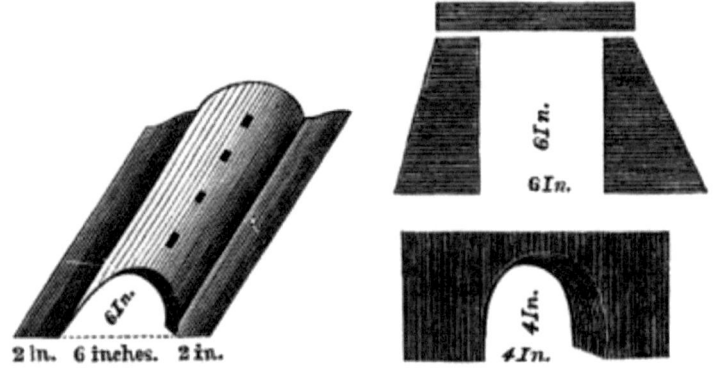

Fig. 26. — Draining-Bricks.

So lately as 1856, tiles were brought from Albany, N. Y., to Exeter, N. H., nearly 300 miles, by railway, at a cost, including freight, of $25 a thousand for two-inch pipes, and it is believed that no tiles were ever made in New Hampshire till the year 1857. These facts will soon become curiosities in agricultural literature, and so are worth preserving. They furnish excuse, too, for what may appear to learned agriculturists an unnecessary particularity in what might seem the well-known facts relative to tile-drainage.

Drain-tiles are made of clay of almost any quality that will make bricks, moulded by a machine into tubes, or into half-tube or horse-shoe forms, usually fourteen inches long before drying, and burnt in a furnace or kiln to be about as hard as what are called hard-burnt bricks. They are usually moulded about half an inch in thickness, varying with the size and form of the tile. The sizes vary from one inch to six inches, and sometimes larger, in the diameter of the bore. The forms are also very various; and as this is one of the most essential matters, [122] as affecting the efficiency, the cost, and the durability of tile-drainage, it will be well to give it critical attention.

## THE FORMS OF TILES.

The simplest, cheapest, and best form of drain-tile is the cylinder, or merely a tube, round outside and with a round bore.

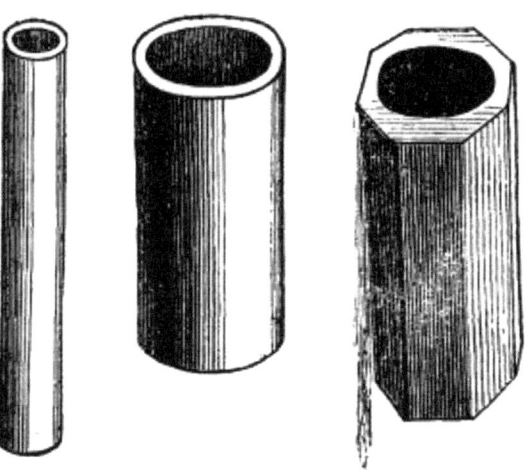

Figs. 27, 28, 29.—Round Pipes.

Tiles of this form, and all others which are tubular, are called *pipes*, in distinction from those with open bottoms, like those of horse-shoe form.

About forty years ago, as Mr. Gisborne informs us, small pipes for land-drainage were used, concurrently, by persons residing in the counties of Lincoln, Oxford, and Kent, who had, probably, no knowledge of each other's operations. Most of those pipes were made with eyelet-holes, to admit the water. Pipes for thorough-draining excited no general attention till they were exhibited by John Read at the show at Derby, in the year 1843. A medal was awarded to the exhibitor. Mr. Parkes was one of the judges, and brought the pipes to the special notice of the council. From this time, inventions and improvements were rapid, and soon, collars were introduced, and the use of improved machines to mould the pipes; [123] and drainage, under the fostering influence of the Royal Agricultural Society, became a subject of general attention through-out the kingdom. The round pipe, or *the pipe*, as it seems, *par excellence*, to be termed by English drainers, though one of the latest, if not the last form of tiles introduced in England, has become alto-gether the most popular among scientific men, and is generally used in all works conducted under the charge of the Land Drainage

Companies. This ought to settle the question for us, when we consider that the immense sum of twenty millions of dollars of public funds has been expended by them, in addition to vast amounts of private funds, and that the highest practical talent of the nation is engaged in the work.

After giving some idea of the various forms of tiles in use, it is, however, proposed to examine the question upon its merits, so that each may judge for himself which is best.

The earliest form of tiles introduced for the purpose of thorough-drainage, was the horse-shoe tile, so called from its shape. The horse-shoe tile has been sometimes used without any sole to form the bottom of the drain, thus leaving the water to run on the ground. There can hardly be a question of the false economy of this mode, for the hardest and most impervious soil softens under the constant action of running water, and then the edges of the tiles must sink, or the bottom of the drain rise, and thus destroy the work.

Various devices have been tried to save the expense of soles, such as providing the edges of the tiles with flanges or using pieces of soles on which to rest the ends of the tiles. They all leave the bottom of the drain unprotected against the wearing action of the water.

Horse-shoe tiles, or "tops and bottoms" as they are called in some counties, are still much used in England; [124] and in personal conversation with farmers there, the writer found a strong opinion expressed in their favor. The advantages claimed for the "tops and bottoms" are, that they lie firmly in place, and that they admit the water more freely than others.

The objections to them are, that they are more expensive than round pipes, and are not so strong, and are not so easily laid, and that they do not discharge water so well as tiles with a round bore. In laying them, they should be made to rest partly upon two adjoining soles, or to break bond, as it is called. The soles are made separate from the tiles, and are merely flat pieces, of sufficient width to support firmly both edges of the tiles. The soles are usually an inch wider than the tiles.

Fig. 30 — Horse-shoe Tiles and Soles.

The above figure represents the horse-shoe tiles and soles properly placed.

As this form of tile has been generally used by the most successful drainers in New York, it may be well to cite the high authority of Mr. Gisborne for the objections which have been suggested. It should be recollected in this connection, that the drainage in this country has been what in England would be called shallow, and that it is too recent to have borne the test of time.

Mr. Gisborne says:

"We shall shock and surprise many of our readers, when we state confidently that, in average soils, and still more in those which are inclined to be tender, horse-shoe tiles form the weakest and most failing conduit which has ever been used for a deep drain. It is so, however; and a little thought, even if we had no experience, will tell us that it must be so.

"A horse-shoe tile, which may be a tolerably secure conduit in a drain of 2 feet, in one of 4 feet becomes an almost certain failure. As [125] to the longitudinal fracture, not only is the tile subject to be broken by one of those slips which are so troublesome in deep draining, and to which the lightly-filled material, even when the drain is completed, offers an imperfect resistance, but the constant pressure together of the sides, even when it does not produce a fracture of the soil, catches hold of the feet of the tile, and breaks it through the crown. When the Regent's Park was first drained, large conduits were in fashion, and they were made circular by placing one horse-shoe tile upon another. It would be difficult to invent a weaker conduit. On re-drainage, innumerable instances were found in which the upper tile was broken through the crown and had dropped into the lower."

Another form of tiles, called *sole-tiles*, or *sole-pipes*, is much used in America, more indeed than any other, except perhaps the horse-shoe tile; probably, because the first manufacturers fancied them the best, and offered no others in the market.

In this form, the sole is solid with the tile. The bottom is flat, but the bore is round, or oval, or egg-shaped, with the small end of the orifice downward.

Fig. 31 — Sole-Tile.

The sole-pipe has considerable advantages theoretically. The opening or bore is of the right shape, the bottom lies fair and firm in place, and the drain, indeed, is perfect, if carefully and properly laid.

The objections to the sole-pipes are, that they are somewhat more expensive than round pipes, and that they require great care in placing them, so as to make the passage even from one pipe to another.

A slight depression of one side of a pipe of this kind, especially if the bore be oval or egg-shaped, throws the water passage out of line. In laying them, the author has taken the precaution to place under each joint a thin piece of wood, such as our honest shoe manufacturers use for [126] stiffening in shoes, to keep the bottoms of the pipes even, at least until the ground has settled compactly, and as much longer as they may escape "decay's effacing finger."

Collars for tiles are used wherever a sudden descent occurs in the course of a drain, or where there is a loose sand or a boggy place, and by many persons they are used in all drains through sandy or gravelly land.

Fig. 32. — Pipes and Collar.

The above figure represents pipe-tiles fitted with collars. Collars are merely short sections of pipes of such size as to fit upon the smaller ones loosely, covering the joint, and holding the ends in place, so that they cannot slip past each other. In very bad places, small pipes may be entirely sheathed in larger ones; and this is advisable in steep descents or flowing sands.

A great advantage in round pipes is, that there is no wrong-side-up to them, and they are, therefore, more readily placed in position than tiles of any other form.

Again: all tiles are more or less warped in drying and burning; and, where it is desired to make perfect work, round pipes may be turned so as to make better joints and a straighter run for the water—which is very important.

If collars are used, there is still less difficulty in adjusting the pipes so as to make the lines straight, and far less danger of obstruction by sand or roots. Indeed, it is believed that no drain can be made more perfect than with round pipes and collars.

As it is believed that few collars have ever yet been used in this country, and the best drainers in England are not agreed as to the necessity of using them, we give the opinions of two or three distinguished gentlemen, in their own language. Mr. Gisborne says:

"We were astounded to find, at the conclusion of Mr. Parkes' Newcastle [127] Lecture, this sentence: 'It may be advisable for me to say, that in clays, and other clean-cutting and firm-bottomed soils, I do not find the collars to be indispensably necessary, although I always prefer their use.' This is a barefaced treachery to pipes, an abandonment of the strongest point in their case—the assured continuity of the conduit. Every one may see how very small a disturbance at their point of junction would dissociate two pipes of one inch diameter. One finds a soft place in the bottom of the drain and dips his nose into it one inch deep, and cocks up his other end. By this simple operation, the continuity of the conduit is twice broken. An inch of lateral motion produces the same effect. Pipes of a larger diameter than two inches are generally laid without collars. This is a practice on which we do not look with much complacency; it is the compromise between cost and security, to which the affairs of men are so often compelled. No doubt, a conduit from three to six inches in diameter is much less subject to a breach in its continuity than one which is smaller; but, when no collars are used, the pipes should be laid with extreme care, and the bed which is prepared for them at the bottom of the drain should be worked to their size and shape with great accuracy.

"To one advantage which is derived from the use of collars we have not yet adverted—the increased facility with which free water existing in the soil can find entrance into the conduit.

"The collar for a one and a half inch pipe has a circumference of nine inches. The whole space between the collar and the pipe, on each side of the collar, is open, and affords no resistance to the entrance of water: while, at the same time, the superincumbent arch of the collar protects the junction of two pipes from the intrusion of particles of soil. We confess to some original misgivings, that a pipe resting only on an inch at each end, and lying hollow, might prove weak, and liable to fracture by weight pressing on it from above; but the fear was illusory. Small particles of soil trickle down the sides of every drain, and the first flow of water will deposit them in the vacant space between the two collars. The bottom, if at all soft, will also swell up into any vacancy. Practically, if you re-open a drain well laid with pipes and collars, you will find them reposing in a beautiful nidus, which, when they are carefully removed, looks exactly as if it had been moulded for them."

As to the danger of breaking the pipes, which might well be apprehended, we found by actual experiment, at the New York Central Park, that a one-inch Albany pipe [128] resting on collars upon a floor, with a bearing at each end of but one inch, would support the weight of a man weighing 160 pounds, standing on one foot on the middle of the pipe.

Mr. Parkes sums up his opinion upon the subject of collars, in these words:

"It may be advisable for me to say, that in clays, and other clean-cutting and firm-bottomed soils, I do not find collars to be at all necessary; but that they are essential in all sandy, loose, and soft strata."

In draining in the neighborhood of trees, collars are also supposed to be of great use in preventing the intrusion of roots into the pipes, although it may be impossible, even in this way, to exclude the roots of water-loving trees.

From the most careful inquiry that the writer was able to make, as to the practice in England, he is satisfied that collars are not generally used there in the drainage of clays, but that the pipes are laid in openings shaped for them at the bottom of the drains, with a tool which forms a groove into which the pipes fall readily into line, and very little seems to be said of collars in the published estimates of the cost of drainage.

On this subject, we have the opinion of Mr. Denton, thus expressed:

"The use of collars is by no means general, although those who have used them speak highly of their advantages. Except in sandy soils, and in those that are subject to sudden alteration of character, in some of the deposits of red sand-stones, and in the clayey subsoils of the Bagshot sand district, for instance, collars are not found to be essential to good drainage. In the north of England they are used but seldom, and, in my opinion, much less than they ought to be; but this opinion, it is right to state, is opposed, in numerous instances of successful drainage, by men of extensive practice; and as every cause of increased outlay is to be avoided, the value of collars, as general appliances, remains an open question. In all the more porous subsoils in which collars have not been used, the more successful drainers increase the size of the pipes in the minor drains to a minimum size of two inches bore."

[129]

*The form of the bore, or water passage*, in tiles, is a point of more importance than at first appears. At one of our colleges, certain plank sewers, in the ordinary square form, were often obstructed by the sediment from the dirty water. "Turn them cornerwise," suggested the professor of Natural Philosophy. It was done, and ever after they kept in order. The pressure of water depends on its height, or head. Everybody knows that six feet of water carries a mill-wheel better than one foot. The same principle operates on a small scale. An inch head of water presses harder than a half inch. The *velocity* of water, again, depends much on its height. Whether there be much or little water passing through a drain, it has manifestly a greater power to make its way, to drive before it sand or other obstructions, when it is heaped up in a round passage, than when wandering over the flat surface of a tile sole. Any one who has observed the discharge of water from flat-bottomed and round tiles, will be satisfied that the quantity of water which is sufficient to run in a rapid stream of a half or quarter inch diameter from a round tile, will lazily creep along the flat bottom of a sole tile, with hardly force sufficient to turn aside a grain of sand, or to bring back to light an enterprising cricket that may have entered on an exploration. On the whole, solid tiles, with flat-bottomed passages, may be set down among the inventions of the adversary. They have not the claims even of the horse-shoe form to respect, because they do not admit water better than round pipes, and are not united by a sole on which the ends of the adjoining tiles rest. They combine the faults of all other forms, with the peculiar virtues of none.

[130]

Fig. 33 — Flat-bottomed Pipe-Tile.

From an English report on the drainage of towns, the following, which illustrates this point, is taken:

"It was found that a large proportion of sewers were constructed with flat bottoms, which, when there was a small discharge, spread the water, increased the friction, retarded the flow, and accumulated deposit. It was ascertained, that by the substitution of circular sewers of the same width, with the same inclination and the same run of water, the amount of deposit was reduced more than one-half."

## THE SIZE OF TILES.

Is a matter of much importance, whether we regard the efficiency and durability of our work, or economy in completing it. The cost of tiles, and the freight of them, increase rapidly with their size, and it is, therefore, well to use the smallest that will effect the object in view. Tiles should be large enough, as a first proposition, to carry off, in a reasonable time, all the surplus water that may fall upon the land. Here, the English rules will not be safe for us; for, although England has many more rainy days than we have, yet we have, in general, a greater fall of rain—more inches of water from the clouds in the year. Instead of their eternal drizzle, we have thunder showers in Summer, and in Spring and Autumn north-east storms, when the windows of heaven are opened, and a deluge, except in duration, bursts upon us. Then, at the North, the Winter snows cover the fields until April, when they suddenly dissolve, often under heavy showers of rain, and planting time is at once upon us. It is desirable that all the snow and rain-water should pass through the soil into the drains, instead of overflowing the surface, so as to save the elements of fertility with which such water abounds, and also to prevent the washing of the soil. We require, then, a greater capacity of drainage, larger tiles, than do the English, for our drains must do a greater work than theirs, and in less time. [131]

There are several other general considerations that should be noticed, before we attempt to define the particular size for any location. Several small drains are usually discharged into one main drain. This main should have sufficient capacity to conduct all the water that may be expected to enter it, and no more. If the small drains overflow it, the main will be liable to be burst, or the land about it filled with water, gushing from it at the joints; especially, if the small drains come down a hill side, so as to give a great pressure, or head of water. On the other hand, if the main be larger than is necessary, there is the useless expense of larger tiles than were required. The capacity of pipes to convey water, depends, other things being equal, upon their size; but here the word size has a meaning which should be kept clearly in mind.

The capacity of round water-pipes is in proportion to the squares of their diameters.

A one-inch pipe carries one inch (circular, not square) of water, but a two-inch pipe carries not two inches only, but twice two, or four inches of water; a three-inch pipe carries three times three, or nine inches; and a four-inch pipe, sixteen inches. Thus we see, that under the same conditions as to fall, directness, smoothness, and the like, a four-inch pipe carries just four times as much water as a two-inch pipe. In fact, it will carry more than this proportion, because *friction*, which is an important element in all such calculations, is greater in proportion to the smaller size of the pipe.

Velocity is another essential element to be noticed in determining the amount of water which may be discharged through a pipe of given diameter. Velocity, again, depends on several conditions. Water runs faster down a steep hill than down a gentle declivity. This is due to the weight of the water, or, in other words, to gravitation, and operates whether the water be at large on [132] the ground, or confined in a pipe, and it operates alike whether the water in a pipe fill its bore or not.

But, again, the velocity of water in a pipe depends on the pressure, or head of water, behind it, and there is, perhaps, no definite limit to the quantity of water that may be forced through a given orifice. More water, for instance, is often forced through the pipe of a fire-engine in full play, in ten minutes, than would run through a pipe of the same diameter, lying nearly level in the ground, in ten hours.

In ordinary aqueducts, for supplying water, and not for drainage, it is desirable to have a high pressure upon the pipes to ensure a rapid flow; but in drainage, a careful distinction must be made between velocity induced by gravitation, and velocity induced by pressure. If induced by the former merely, the pipe through which the water is swiftly running, if not quite full, may still receive water at every joint, while, if the velocity be induced by pressure, the pipe must be already full. It can then receive no more, and must lose water at the joints, and wet the land through which it passes, instead of draining it.

So that although we should find that the mains might carry a vast quantity of water admitted by minor drains from high elevations, yet we should bear in mind, that drains when full can perform no ordinary office of drainage. If there is more than the pressure of four feet head of water behind; the pipes, if they passed through a pond of water, at four feet deep, must lose and not receive water at the joints.

The capacity of a pipe to convey water depends, then, not only on its size, but on its inclination or fall—a pipe running down a considerable descent having much greater capacity than one of the same size lying nearly level. This fact should be borne in mind even in laying single drains; for it is obvious that if the drain lie along a sandy [133] plain, for instance, extending down a springy hill-side, and then, as is usually the case, along a lower plain again, to its outlet at some stream, it may collect as much water as will fill it before it reaches the lower level. Its stream rushes swiftly down the descent, and when it reaches the plain, there is not sufficient fall to carry it away by its natural gravitation. It will still rush onward to its outlet, urged by the pressure from behind; but, with such pressure, it will, as we have seen, instead of draining the land, suffuse it with water.

## FRICTION,

as has already been suggested, is an element that much interferes with exact calculations as to the relative capacity of water-pipes of various dimensions, and this depends upon several circumstances, such as smoothness, and exactness of form, and directness. The smoother, the more regular in form, and the straighter the drain, the more water will it convey. Thus, in some recent English experiments,

"it was found that, with pipes of the same diameter, exactitude of form was of more importance than smoothness of surface; that glass pipes, which had a wavy surface, discharged less water, at the same inclinations, than Staffordshire stone-ware clay pipes, which were of perfectly exact construction. By passing pipes of the same clay—the common red clay—under a second pressure, obtained by a machine at an extra expense of about eighteen pence per thousand, whilst the pipe was half dry, very superior exactitude of form was obtained, and by means of this exactitude, and with nearly the same diameters, an increased discharge of water of one-fourth was effected within the same time."

So all sudden turns or angles increase friction and retard velocity, and thus lessen the capacity of the drain—a topic which may be more properly considered under the head of the junction of drains.

"On a large scale, it was found that when equal quantities of water were running direct, at a rate of 90 seconds, with a turn at right-angles, [134] the discharge was only effected in 140 seconds; whilst, with a turn or junction with a gentle curve, the discharge was effected in 100 seconds."

We are indebted to Messrs. Shedd & Edson for the following valuable tables showing the capacity of water-pipes, with the accompanying suggestions:

## "DISCHARGE OF WATER THROUGH PIPES.

"The following tables of discharge are founded on the experiments made by Mr. Smeaton, and have been compared with those by Henry Law, and with the rules of Weisbach and D'Aubuisson. The conditions under which such experiments are made may be so essentially different in each case, that few experiments give results coincident with each other, or with the deductions of theory: and in applying these tables to practice, it is quite likely that the discharge of a pipe of a certain area, at a certain inclination, may be quite unlike the discharge found to be due to those conditions by this table, and that difference may be owing partly to greater or less roughness on the inside of the pipe, unequal flow of water through the joints into the pipe, crookedness of the pipes, want of accuracy in their being placed, so that the fall may not be uniform throughout, or the ends of the pipes may be shoved a little to one side, so that the continuity of the channel is partially broken; and, indeed, from various other causes, all of which may occur in any practical case, unless great care is taken to avoid it, and some of which may occur in almost any case.

"We have endeavored to so construct the tables that, in the ordinary practice of draining, the discharge given may approximate to the truth for a well laid drain, subject even to considerable friction. The experiments of Mr. Smeaton, which we have adopted as the basis of these tables, gave a less quantity discharged, under certain conditions, than given under similar conditions by other tables. This result is probably due to a greater amount of friction in the pipes used by Smeaton. The curves of friction resemble, very nearly, parabolic curves, but are not quite so sharp near the origin.

"We propose, during the coming season, to institute some careful experiments, to ascertain the friction due to our own drain-pipe. Water can get into the drain-pipe very freely at the joints, as may be seen by a simple calculation. It is impossible to place the ends so closely together, in laying, as to make a tight joint on account of roughness in the clay, twisting in burning, &c.; and the opening thus made will usually average about one-tenth of an inch on the whole circumference, [135] which is, on the inside of a two-inch pipe, six inches—making six-tenths of a square inch opening for the entrance of water at each joint.

"In a lateral drain 200 feet long, the pipes being thirteen inches long, there will be 184 joints, each joint having an opening of six-tenth square inch area; in 184 joints there is an aggregate area of 110 square inches; the area of the opening at the end of a two-inch pipe is about three inches; 110 square inches inlet to three inches outlet; thirty-seven times as much water can flow in as can flow out. There is, then, no need for the water to go through the pores of the pipe; and the fact is, we think, quite fortunate, for the passage of water through the pores would in no case be sufficient to benefit the land to much extent. We tried an experiment, by stopping one end of an ordinary drain-pipe and filling it with water. At the end of sixty-five hours, water still stood in the pipe three-fourths of an inch deep. About half the water first put into the pipe had run out at the end of twenty-four hours. If the pipe was stopped at both ends and plunged four feet deep in water, it would undoubtedly fill in a short time; but such a test is an unfair one, for no drain could be doing service, over which water could collect to the depth of four feet."

1½-inch drain-pipe.
Area: 1.76709 inches.

| Fall in 100 feet. | Velocity per second in feet. | Discharge in gallons in 24 hours. | Fall in 100 feet. | Velocity per second in feet. | Discharge in gallons in 24 hours. |
|---|---|---|---|---|---|

| ft. in. | | | ft. in. | | |
|---|---|---|---|---|---|
| 0.3 | 0.71 | 5630.87 | 5.3 | 3.75 | 29704.51 |
| 0.6 | 1.04 | 8248.03 | 5.6 | 3.84 | 30454.28 |
| 0.9 | 1.29 | 10230.73 | 5.9 | 3.93 | 31168.06 |
| 1.0 | 1.52 | 12054.81 | 6.0 | 4.00 | 31723.21 |
| 1.3 | 1.74 | 13799.59 | 6.3 | 4.10 | 32516.36 |
| 1.6 | 1.91 | 15147.83 | 6.6 | 4.18 | 33150.76 |
| 1.9 | 2.10 | 16654.68 | 6.9 | 4.25 | 33705.91 |
| 2.0 | 2.26 | 17923.61 | 7.0 | 4.33 | 34340.38 |
| 2.3 | 2.41 | 19113.23 | 7.3 | 4.41 | 34974.85 |
| 2.6 | 2.56 | 20302.86 | 7.6 | 4.49 | 35609.30 |
| 2.9 | 2.69 | 21333.86 | 7.9 | 4.56 | 36154.45 |
| 3.0 | 2.83 | 22444.17 | 8.0 | 4.65 | 36878.23 |
| 3.3 | 2.94 | 23150.71 | 8.3 | 4.71 | 37354.08 |
| 3.6 | 3.06 | 24268.25 | 8.6 | 4.79 | 37988.55 |
| 3.9 | 3.16 | 25061.34 | 8.9 | 4.85 | 38464.40 |
| 4.0 | 3.28 | 26013.03 | 9.0 | 4.91 | 38940.25 |
| 4.3 | 3.38 | 26806.11 | 9.3 | 4.98 | 39495.39 |
| 4.6 | 3.46 | 27440.58 | 9.6 | 5.04 | 39971.24 |
| 4.9 | 3.56 | 28233.66 | 9.9 | 5.10 | 40447.10 |
| 5.0 | 3.65 | 28947.43 | 10.0 | 5.16 | 40922.93 |

| [136] 2-inch drain-pipe. | | | 3-inch drain-pipe. | | |
|---|---|---|---|---|---|
| Fall in 100 feet. | Velocity per second in feet. | Discharge in gallons in 24 hours. | Fall in 100 feet. | Velocity per second in feet. | Discharge in gallons in 24 hours. |
| ft. in. | | | ft. in. | | |
| 0.3 | 0.79 | 10575.4 | 0.3 | 0.90 | 24687.2 |
| 0.6 | 1.16 | 15528.4 | 0.6 | 1.33 | 36482.2 |
| 0.9 | 1.50 | 20079.9 | 0.9 | 1.66 | 45534.2 |
| 1.0 | 1.71 | 22891.1 | 1.0 | 1.94 | 53214.7 |
| 1.3 | 1.94 | 25970.0 | 1.3 | 2.19 | 60072.2 |
| 1.6 | 2.16 | 28915.1 | 1.6 | 2.43 | 66655.5 |
| 1.9 | 2.35 | 31458.5 | 1.9 | 2.63 | 72141.5 |
| 2.0 | 2.53 | 33868.1 | 2.0 | 2.83 | 77627.6 |
| 2.3 | 2.69 | 36009.9 | 2.3 | 3.00 | 82290.7 |
| 2.6 | 2.83 | 37884.0 | 2.6 | 3.16 | 86679.6 |
| 2.9 | 2.97 | 39758.2 | 2.9 | 3.31 | 90794.1 |
| 3.0 | 3.11 | 41632.4 | 3.0 | 3.47 | 95182.9 |
| 3.3 | 3.24 | 43372.6 | 3.3 | 3.60 | 98748.9 |
| 3.6 | 3.36 | 44979.0 | 3.6 | 3.74 | 102589.1 |
| 3.9 | 3.48 | 46585.4 | 3.9 | 3.87 | 106155.0 |
| 4.0 | 3.59 | 48057.9 | 4.0 | 3.99 | 109446.7 |
| 4.3 | 3.70 | 49530.5 | 4.3 | 4.11 | 112738.3 |
| 4.6 | 3.80 | 50869.1 | 4.6 | 4.23 | 116029.9 |

| | | | | | |
|---|---|---|---|---|---|
| 4.9 | 3.91 | 52341.6 | 4.9 | 4.34 | 119047.3 |
| 5.0 | 4.02 | 53814.1 | 5.0 | 4.46 | 122338.9 |
| 5.3 | 4.11 | 55018.9 | 5.3 | 4.57 | 125356.2 |
| 5.6 | 4.22 | 56491.5 | 5.6 | 4.68 | 128373.5 |
| 5.9 | 4.31 | 57696.3 | 5.9 | 4.78 | 131116.6 |
| 6.0 | 4.40 | 58901.1 | 6.0 | 4.89 | 134133.9 |
| 6.3 | 4.49 | 60105.9 | 6.3 | 4.98 | 136602.6 |
| 6.6 | 4.58 | 61309.7 | 6.6 | 5.08 | 139345.6 |
| 6.9 | 4.66 | 62381.6 | 6.9 | 5.18 | 142088.7 |
| 7.0 | 4.74 | 63452.5 | 7.0 | 5.27 | 144557.4 |
| 7.3 | 4.83 | 64667.3 | 7.3 | 5.37 | 147306.4 |
| 7.6 | 4.91 | 65728.3 | 7.6 | 5.46 | 150069.1 |
| 7.9 | 4.99 | 66799.2 | 7.9 | 5.55 | 152237.8 |
| 8.0 | 5.07 | 67870.1 | 8.0 | 5.64 | 154706.6 |
| 8.3 | 5.15 | 68941.0 | 8.3 | 5.73 | 157175.3 |
| 8.6 | 5.23 | 70011.9 | 8.6 | 5.82 | 159644.0 |
| 8.9 | 5.31 | 71082.8 | 8.9 | 5.91 | 162112.7 |
| 9.0 | 5.38 | 72019.9 | 9.0 | 5.99 | 164313.2 |
| 9.3 | 5.46 | 73090.9 | 9.3 | 6.07 | 166501.6 |
| 9.6 | 5.53 | 74027.9 | 9.6 | 6.16 | 168970.3 |
| 9.9 | 5.60 | 74965.0 | 9.9 | 6.24 | 171164.7 |
| 10.0 | 5.67 | 75902.0 | 10.0 | 6.32 | 173359.1 |

| [137] 4-inch drain-pipe. | | | 5-inch drain-pipe. | | |
|---|---|---|---|---|---|
| Fall in 100 feet. | Velocity per second in feet. | Discharge in gallons in 24 hours. | Fall in 100 feet. | Velocity per second in feet. | Discharge in gallons in 24 hours. |
| ft. in. | | | ft. in. | | |
| 0.3 | 1.08 | 43697.6 | 0.3 | 1.13 | 99584.2 |
| 0.6 | 1.50 | 60691.2 | 0.6 | 1.57 | 138362.4 |
| 0.9 | 1.83 | 74043.2 | 0.9 | 1.90 | 167442.6 |
| 1.0 | 2.13 | 86181.4 | 1.0 | 2.20 | 193881.0 |
| 1.3 | 2.38 | 96296.6 | 1.3 | 2.45 | 215912.9 |
| 1.6 | 2.61 | 105602.6 | 1.6 | 2.70 | 237944.9 |
| 1.9 | 2.81 | 113694.8 | 1.9 | 2.90 | 255569.5 |
| 2.0 | 3.00 | 121382.3 | 2.0 | 3.10 | 273195.9 |
| 2.3 | 3.19 | 129089.9 | 2.3 | 3.29 | 289940.1 |
| 2.6 | 3.36 | 135948.2 | 2.6 | 3.46 | 304921.9 |
| 2.9 | 3.53 | 142826.5 | 2.9 | 3.64 | 320784.9 |
| 3.0 | 3.68 | 148895.7 | 3.0 | 3.80 | 334885.4 |
| 3.3 | 3.82 | 154560.2 | 3.3 | 3.96 | 348974.8 |
| 3.6 | 3.96 | 160224.7 | 3.6 | 4.11 | 362204.9 |
| 3.9 | 4.10 | 165889.2 | 3.9 | 4.26 | 375424.1 |
| 4.0 | 4.24 | 171553.7 | 4.0 | 4.40 | 387762.1 |
| 4.3 | 4.37 | 176813.6 | 4.3 | 4.52 | 398337.5 |
| 4.6 | 4.50 | 182073.5 | 4.6 | 4.66 | 410675.3 |

| | | | | | |
|---|---|---|---|---|---|
| 4.9 | 4.62 | 186928.3 | 4.9 | 4.78 | 421250.6 |
| 5.0 | 4.75 | 192188.7 | 5.0 | 4.90 | 430825.0 |
| 5.3 | 4.86 | 196639.4 | 5.3 | 5.02 | 442401.3 |
| 5.6 | 4.97 | 201090.1 | 5.6 | 5.14 | 452976.6 |
| 5.9 | 5.09 | 205945.3 | 5.9 | 5.25 | 462670.6 |
| 6.0 | 5.20 | 210396.0 | 6.0 | 5.37 | 473246.0 |
| 6.3 | 5.30 | 214442.1 | 6.3 | 5.49 | 483820.4 |
| 6.6 | 5.41 | 218892.8 | 6.6 | 5.60 | 493514.6 |
| 6.9 | 5.51 | 222938.8 | 6.9 | 5.70 | 502327.4 |
| 7.0 | 5.61 | 226984.9 | 7.0 | 5.80 | 511140.2 |
| 7.3 | 5.71 | 231031.0 | 7.3 | 5.90 | 520052.0 |
| 7.6 | 5.81 | 235077.1 | 7.6 | 6.00 | 528766.5 |
| 7.9 | 5.91 | 239123.2 | 7.9 | 6.10 | 537578.7 |
| 8.0 | 6.01 | 243169.2 | 8.0 | 6.20 | 546391.5 |
| 8.3 | 6.10 | 246810.7 | 8.3 | 6.30 | 555204.5 |
| 8.6 | 6.19 | 250452.2 | 8.6 | 6.40 | 564017.0 |
| 8.9 | 6.28 | 255493.7 | 8.9 | 6.49 | 571948.0 |
| 9.0 | 6.37 | 257735.2 | 9.0 | 6.58 | 579880.0 |
| 9.3 | 6.45 | 260971.9 | 9.3 | 6.66 | 586930.2 |
| 9.6 | 6.54 | 264603.1 | 9.6 | 6.75 | 594861.4 |
| 9.9 | 6.63 | 268254.9 | 9.9 | 6.84 | 602793.2 |
| 10.0 | 6.71 | 271491.8 | 10.0 | 6.93 | 610723.8 |

[138] 8-inch drain-pipe.
Area: 50.2640 inches.

| Fall in 100 feet. | Velocity per second in feet. | Discharge in gallons in 24 hours. | Fall in 100 feet. | Velocity per second in feet. | Discharge in gallons in 24 hours. |
|---|---|---|---|---|---|
| ft. in. | | | ft. in. | | |
| 0.3 | 1.23 | 277487.7 | 5.3 | 5.35 | 1206959.3 |
| 0.6 | 1.65 | 372239.7 | 5.6 | 5.47 | 1234031.3 |
| 0.9 | 2.01 | 453455.7 | 5.9 | 5.59 | 1261103.3 |
| 1.0 | 2.33 | 525647.7 | 6.0 | 5.71 | 1288175.3 |
| 1.3 | 2.60 | 586559.7 | 6.3 | 5.83 | 1315247.3 |
| 1.6 | 2.85 | 642959.6 | 6.6 | 5.95 | 1343838.9 |
| 1.9 | 3.08 | 694847.6 | 6.9 | 6.07 | 1369391.3 |
| 2.0 | 3.30 | 744479.7 | 7.0 | 6.17 | 1391951.2 |
| 2.3 | 3.50 | 789599.6 | 7.3 | 6.27 | 1414531.1 |
| 2.6 | 3.70 | 844719.7 | 7.6 | 6.39 | 1441583.2 |
| 2.9 | 3.89 | 877583.5 | 7.9 | 6.50 | 1466399.3 |
| 3.0 | 4.05 | 913679.5 | 8.0 | 6.60 | 1488959.2 |
| 3.3 | 4.21 | 949775.6 | 8.3 | 6.70 | 1511539.1 |
| 3.6 | 4.37 | 971658.7 | 8.6 | 6.80 | 1534099.0 |
| 3.9 | 4.53 | 920447.4 | 8.9 | 6.90 | 1556658.9 |
| 4.0 | 4.67 | 1055551.4 | 9.0 | 7.00 | 1579199.3 |
| 4.3 | 4.81 | 1086135.4 | 9.3 | 7.10 | 1601759.2 |

| 4.6 | 4.95 | 1116718.7 | 9.6 | 7.20 | 1624319.1 |
| 4.9 | 5.08 | 1146047.4 | 9.9 | 7.29 | 1644622.1 |
| 5.0 | 5.22 | 1177631.3 | 10.0 | 7.38 | 1664927.1 |

## HOW WATER ENTERS THE TILES.

How water enters the tiles, is a question which all persons unaccustomed to the operation of tile-draining usually ask at the outset. In brief, it may be answered, that it enters both at the joints and through the pores of the burnt clay, but mostly at the joints.

Mr. Parkes expresses the opinion, based upon careful observation, that five hundred times as much water enters at the crevices as through the pores of the tiles! If this be so, we may as well, for all practical purposes, regard the water as all entering at the joints. In several experiments which we have attempted, we have found the quantity of water that enters through the pores to be quite too small to be of much practical account.

Tiles differ so much in porosity, that it is difficult to [139] make experiments that can be satisfactory — soft-burnt tiles being, like pale bricks, quite pervious, and hard-burnt tiles being nearly or quite impervious. The amount of pressure upon the clay in moulding also affects the density and porosity of tiles.

Water should enter at the bottom of the tiles, and not at the top. It is a well-known fact in draining, that the deepest drain flows first and longest. A familiar illustration will make this point evident. If a cask or deep box be filled with sand, with one hole near the bottom and another half way to the top, these holes will represent the tiles in a drain. If water be poured into the sand, it will pass downward to the bottom of the vessel, and will not flow out of either hole till the sand be saturated up to the lower hole, and then it will flow out there. If, now, water be poured in faster than the lower hole can discharge it, the vessel will be filled higher, till it will run out at both holes. It is manifest, however, that it will first cease to flow from the upper orifice. There is in the soil a line of water, called the "water-line," or "water-table;" and this, in drained land, is at about the level of the bottom of the tiles. As the rain falls it descends, as in the vessel; and as the water rises, it enters the tiles at the bottom, and never at the top, unless there is more than can pass out of the soil by the lower openings (the crevices and pores) into the tiles. It is well always to interrupt the direct descent of water by percolation from the surface to the top of the tiles, because, in passing so short a distance in the soil, the water is not sufficiently filtered, especially in soil so recently disturbed, but is likely to carry with it not only valuable elements of fertility, but also particles of sand, which may obstruct the drain. This is prevented by placing above the tiles (after they are covered a few inches with gravel, sand, or other porous soil) compact clay, if convenient. If not, a furrow each side of the [140] drain, or a heaping-up of the soil over the drain, when finished, will turn aside the surface-water, and prevent such injury.

In the estimates as to the area of the openings between pipes, it should be considered that the spaces between the pipes are not, in fact, clean openings of one-tenth of an inch, but are partially closed by earthy particles, and that water enters them by no means as rapidly as it would enter the clean pipes before they are covered. Although the rain-fall in England is much less in quantity and much more regular than in this country, yet it is believed that the use of two-inch pipes will be found abundantly sufficient for the admission and conveyance of any quantity of water that it may be necessary to carry off by drainage in common soils. In extraordinary cases, as where the land drained is a swamp, or reservoir for water which falls on the hills around, larger pipes must be used.

In many places in England "tops and bottoms," or horse-shoe tiles, are still preferred by farmers, upon the idea that they admit the water more readily; but their use is continued only by those who have never made trial of pipes. No scientific drainer uses any but pipes in England, and the million of acres well drained with them, is pretty good evidence of their sufficiency. In this country, horse-shoe tiles have been much used in Western New York, and have been found to answer a good purpose; and so it may be said of the sole-pipes. Indeed, it is believed that no instance is to be found on record in America of the failure of tile drains, from the inability of the water to gain admission at the joints.

It may be interesting in this connection to state, that water is 815 times heavier than air. Here is a drain at four feet depth in the ground, filled only with air, and open at the end so that the air can go out. Above this open space is four feet of earth saturated with water. What is the pressure of the water upon the tiles? [141]

Mr. Thomas Arkell, in a communication to the Society of Arts, in England, says—

"The pressure due to a head of water four or five feet, may be imagined from the force with which water will come through the crevices of a hatch with that depth of water above it. Now, there is the same pressure of water to enter the vacuum in the pipe-drain as there is against the hatches, supposing the land to be full of water to the surface."

It is difficult to demonstrate the truth of this theory; but the same opinion has been expressed to the writer by persons of learning and of practical skill, based upon observations as to the entrance of water into gas pipes, from which it is almost, if not quite, impossible to exclude it by the most perfect joints in iron pipes. Whatever be the theory as to pressure, or the difficulties as to the water percolating through compact soils to the tiles, there will be no doubt left on the mind of any one, after one experiment tried in the field, that, in common cases, all the surplus water that reaches the tiles is freely admitted. A gentleman, who has commenced draining his farm, recently, in New Hampshire, expressed to the author his opinion, that tiles in his land admitted the water as freely as a hole of a similar size to the bore of the tile would admit it, if it could be kept open through the soil without the tile.

## DURABILITY OF TILE DRAINS.

How long will they last? This is the first and most important question. Men, who have commenced with open ditches, and, having become disgusted with the deformity, the inconvenience, and the inefficiency of them, have then tried bushes, and boards, and turf, and found them, too, perishable; and again have used stones, and after a time seen them fail, through obstructions caused by moles or frost—these men have the right to a well-considered answer to this question. [142]

The foolish fellow in the Greek Reader, who, having heard that a crow would live a hundred years, purchased one to verify the saying, probably did not live long enough to ascertain that it was true. How long a properly laid tile-drain of hard-burnt tiles will endure, has not been definitely ascertained, but it is believed that it will outlast the life of him who lays it.

No tiles have been long enough laid in the United States to test this question by experience, and in England no further result seems to have been arrived at, than that the work is a *permanent* improvement.

In another part of this treatise, may be found some account of Land Drainage Companies, and of Government loans in aid of improvements by drainage in Great Britain. One of these acts provides for a charge on the land for such improvements, to be paid in full in fifty years. That is to say, the expense of the drainage is an incumbrance like a mortgage on the land, at a certain rate of interest, and the tenant or occupant of the land, each year pays the interest and enough more to discharge the debt in just fifty years. Thus, it is assumed by the Government, that the improvement will last fifty years in its full operation, because the last year of the fifty pays precisely the same as every other year.

It may therefore be considered as the settled conviction of all branches of the British government, and of all the best-informed, practical land-drainers in that country, that TILE-DRAINAGE WILL ENDURE FIFTY YEARS AT LEAST, if properly executed.

This is long enough to satisfy any American; for the migratory habits of our citizens, and the constant changes of cultivated fields into village and city lots, prevent our imagination even conceiving the idea that we or our posterity can remain for half a century upon the same farm.

It is much easier, however, to lay tile-drains so that [143] they will not be of use half of fifty years, than to make them permanent in their effect. Tile-drainage, it cannot be too much enforced, is an operation requiring great care and considerable skill — altogether more care and skill than our common laborers, or even most of our farmers, are accustomed to exercise in their farm operations.

A blunder in draining, like the blunder of a physician, may be soon concealed by the grass that grows over it, but can never be corrected. Drainage is a new art in this country, and tile-making is a new art. Without good, hard-burnt tiles, no care or skill can make permanent work.

Tile-drainage will endure so long as the tiles last, if the work be properly done.

There is no reason why a tile should not last in the ground as long as a brick will last. Bricks will fall to pieces in the ground in a very short time if not hard-burnt, while hard-burnt bricks of good clay will last as long as granite.

Tiles must be hard-burnt in order to endure. But this is not all. Drains fail from various other causes than the crumbling of the tiles. They are frequently obstructed by mice, moles, frogs, and vermin of all kinds, if not protected at the outlet. They are often destroyed by the treading of cattle, and by the deposit of mud at the outlet, through insufficient care. They are liable to be filled with sand, through want of care in protecting the joints in laying, and through want of collars, and other means of keeping them in line. They are liable, too, to fill up by deposits of sand and the like, by being laid lower in some places than the parts nearer the outlet, so that the slack places catch and retain whatever is brought down, till the pipe is filled.

Frost is an enemy which in this country we have to [144] contend with, more than in any other, where tile-drainage has been much practiced.

Upon all these points, remarks will be found under the appropriate heads; and these suggestions are repeated here, because we know that haste and want of skill are likely to do much injury to the cause which we advocate. Any work that requires only energy and progress, is safe in American hands; but cautious and slow operations are by no means to their taste.

Dickens says, that on railways and coaches, wherever in England they say, "All right," the Americans use, instead, the phrase, "Go ahead." In tile-drainage, the motto, "All right," will be found far more safe than the motto, "Go ahead."

Instances are given in England of drains laid with handmade tiles, which have operated well for thirty years, and have not yet failed.

Mr. Parkes informs us: "That, about 1804, pipe-tiles made tapering, with one end entering the other, and two inches in the smallest point, were laid down in the park now possessed by Sir Thomas Whichcote, Aswarby, Lincolnshire, and that they still act well."

Stephens gives the following instance of the durability of bricks used in draining:

"Of the durability of common brick, when used in drains, there is a remarkable instance mentioned by Mr. George Guthrie, factor to the Earl of Stair or Calhoun, Wigtonshire. In the execution of modern draining on that estate, some brick-drains, on being intersected, emitted water very freely. According to documents which refer to these drains, it appears that they had been formed by the celebrated Marshal, Earl Stair, *upwards of a hundred years ago*. They were found between the vegetable mould and the clay upon which it rested, between the 'wet and the dry,' as the country phrase has it, and about thirty-one inches below the surface. They presented two forms—one consisting of two bricks set asunder on edge, and the other two laid lengthways across them, leaving between them an opening of four inches square for water, [145] but having no soles. The bricks had not sunk in the least through the sandy clay bottom upon which they rested, as they were three inches broad. The other form was of two bricks laid side by side, as a sole, with two others built or laid on each other, at both sides, upon the solid ground, and covered with flat stones, the building being packed on each side of the drain with broken bricks."

In our chapter upon the "Obstruction of Drains," the various causes which operate against the permanency of drains, are more fully considered.

# CHAPTER VII [146]
# DIRECTION, DISTANCE, AND DEPTH OF DRAINS.

Direction of Drains.—Whence comes the Water?—Inclination of Strata.—Drains across the Slope let Water out as well as Receive it.—Defence against Water from Higher Land.—Open Ditches.—Headers.—Silt-basins.

Distance of Drains.—Depends on Soil, Depth, Climate, Prices, System.—Conclusions as to Distance.

Depth of Drains.—Greatly Increases Cost.—Shallow Drains first tried in England.—10,000 Miles of Shallow Drains laid in Scotland by way of Education.—Drains must be below Subsoil plow, and Frost.—Effect of Frost on Tiles and Aqueducts.

## DIRECTION OF DRAINS.

Whether drains should run up and down the slope of the hill, or directly across it, or in a diagonal line as a compromise between the first two, are questions which beginners in the art and mystery of drainage usually discuss with great zeal. It seems so plain to one man, at the first glance, that, in order to catch the water that is running down under the soil upon the subsoil, from the top of the hill to the bottom, you must cut a ditch across the current, that he sees no occasion to examine the question farther. Another, whose idea is, to catch the water in his drain before it rises to the surface, as it is passing up from below or running along on the subsoil, and keep it from rising higher than the bottom of his ditch, thinks it quite as obvious that the drains should run up and down the slope, that the water, once entering, may remain in the drain, going directly down hill to the outlet. A third hits on the Keythorpe system, and regarding the water as [147] flowing down the slope, under the soil, in certain natural channels in the subsoil, fancies they may best be cut off by drains, in the nature of mains, running diagonally across the slope.

These different ideas of men, if examined, will be found to result mainly from their different notions of the underground circulation of water. In considering the Theory of Moisture, an attempt was made to suggest the different causes of the wetness of land.

To drain land effectually, we must have a correct idea of the sources of the water that makes the particular field too wet; whether it falls from the clouds directly upon it; or whether it falls on land situated above it and sloping towards it, so that the water runs down, as upon a roof, from other fields or slopes to our own; or whether it gushes up in springs which find vent in particular spots, and so is diffused through the soil.

If we have only to take care of the water that falls on our own field, from the clouds, that is quite a different matter from draining the whole adjoining region, and requires a different mode of operation. If your field is in the middle, or at the foot, of an undrained slope, from which the water runs on the surface over your land, or soaks through it toward some stream or swamp below, provision must be made not only for drainage of your own field, but also for partial drainage of your neighbor's above, or at least for defence against his surplus of water.

The first, and leading idea to be kept in mind, as governing this question of the direction of drains, is the simple fact that *water runs down hill*; or, to express the fact more scientifically, water constantly seeks a lower level by the force of gravitation, and the whole object of drains is to open lower and still lower passages, into which the water may fall lower and lower until it is discharged from our field at a safe depth. [148]

Water goes down, then, by its own weight, unless there is something through which it cannot readily pass, to bring it out at the surface. It will go into the drains, only because they are lower than the land drained. It will never go *upward* to find a drain, and it will go toward a drain the more readily, in proportion as the descent is more steep toward it.

To decide properly what direction a drain should have, it is necessary, then, to have a definite and a correct idea as to what office the drain is to perform, what water is to fall into it, what land it is to drain.

Suppose the general plan to be, to lay drains forty feet apart, and four feet deep over the field, and the question now to be determined, as to the *direction*, whether across, or up and down the slope, there being fall enough to render either course practicable. The first point of inquiry is, what is expected of each drain? How much and what land should it drain? The general answer must be, forty feet breadth, either up and down the slope, or across it; according to the direction. But we must be more definite in our inquiry than even this. From *what* forty feet of land will the water fall into the drain? Obviously, from some land in which the water is higher than the bottom of the drain.

If, then, the drain run directly *across* the slope, most of the water that can fall into it, must come from the forty feet breadth of land between the drain in question, and the drain next above it. If the water were falling on an impervious surface, it would all run according to the slope of the surface, in which case, by the way, no drains but those across, could catch any of it except what fell upon the drains. But the whole theory of drainage is otherwise, and is based on the idea that we change the course of the underground flow, by drawing out the water [149] at given points by our drains; or, in other words, that "the water seeks the lowest level in all directions."

Upon the best view the writer has been able to take of the two systems as to the direction of drains, there is but a very small advantage in theory in favor of either over the other, in soil which is homogeneous. But it must be borne in mind that homogeneous soil is rather the exception in nature than the rule.

Without undertaking to advance or defend any peculiar geological views of the structure of the earth, or of the depositions or formations that compose its surface, it may be said, that very often the first four feet of subsoil is composed of strata, or layers of earth of varying porosity.

Beneath sand will be found a stratum of clay, or of compact or cemented gravel, and frequently these strata are numerous and thin. Indeed, if there be not some stratum below the soil, which impedes the passage of water, it would pass downward, and the land would need no artificial drainage. Quite often it will be found that the dip or inclination of the various strata below the soil is different from that of the surface.

The surface may have a considerable slope, while the lower strata lie nearly level, as if they had been cut through by artificial grading.

The following figure from the Cyclopedia of Agriculture, with the explanation, fully illustrates this idea.

"In many subsoils there are thin partings, or layers, of porous materials, interspersed between the strata, which, although not of sufficient capacity to give rise to actual springs, yet exude sufficient water to indicate their presence. These partings occasionally crop out, and give rise to those damp spots, which are to be seen diversifying the surface of fields, when the drying breezes of Spring have begun to act upon them. In the following cut, the light lines represent such partings.

"Now, it will be evident, in draining such land, that if the drains be disposed in a direction transverse or oblique to the slope, it will often [150] happen that the drains, no matter how skillfully planned, will not reach these partings at all, as at A. In this case, the water will continue to flow on in its accustomed channel, and discharge its waters at B.

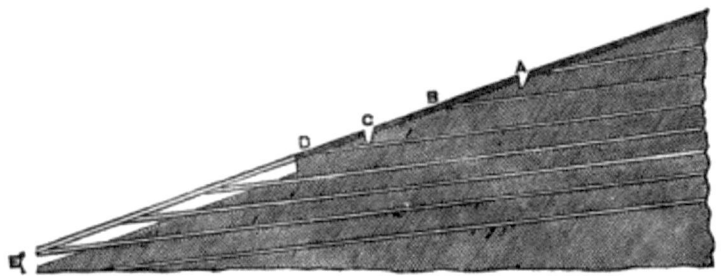

Fig. 34 — Drains across the Slope.

"But again, even though it does reach these partings, as at C, a considerable portion of water will escape from the drain itself, and flow to the *lower level* of its old point of discharge at D. Whereas, a drain cut in the line of the slope, as from D to E, intersects all these partings, and furnishes an outlet to them at a lower level than their old ones."

These reasons are, it is true, applicable only to land of peculiar structure; but there are reasons for selecting the line of greatest fall for the direction of drains which are applicable to all lands alike.

"The line of the greatest fall is the only line in which a drain is relatively lower than the land on either side of it." Whether we regard the surplus water as having recently fallen upon the field, and as being stopped near the surface by an impervious stratum, or as brought down on these strata from above, we have it to be disposed of as it rests upon this stratum, and is borne out by it to the surface.

If there is a decided dip, or inclination, of this stratum outward down the slope, it is manifest that the water cannot pass backward to a cross drain higher up the slope. The course of the water must be downward upon the stratum on which it lies, and so all between two cross [151] drains must pass to the lower one. The upper drain could take very little, if any, and the greater the inclination of this stratum, the less could flow backward.

But in such case a drain down the slope gives to the water borne up by these strata, an outlet of the depth of the drain. If the drain be four feet deep, it cuts the water-bearing strata each at that depth, and takes off the water.

In these cases, the different layers of clay or other impervious "partings," are like the steps of a huge stairway, with the soil filling them up to a regular grade. The ditch cuts through these steps, letting the water that rests on them fall off at the ends, instead of running over the edges. Drains across the slope have been significantly termed "mere catch-waters."

If we wish to use water to irrigate lands, we carefully conduct it along the surface across the slope, allowing it to flow over and to soak through the soil. If we desire to carry the same water off the field as speedily as possible, we should carry our surface ditch directly down the slope.

Now, looking at the operation of drains across the slope, and supposing that each drain is draining the breadth next above it, we will suppose the drain to be running full of water. What is there to prevent the water from passing out of that drain in its progress, at every point of the tiles, and so saturating the breadth below it?

Drainpipes afford the same facility for water to soak out at the lower side, as to enter on the upper, and there is the same law of gravitation to operate in each case. Mr. Denton gives instances in which he has observed, where drains were carried across the slope, in Warwickshire, lines of moisture at a regular distance below the drains. He could ascertain, he says, the depth of the drain itself, by taking the difference of height between the line of the [152] drain at the surface, and that of the line of moisture beneath it. He says again:

"I recently had an opportunity, in Scotland, of gauging the quantity of water traveling along an important drain carried obliquely across the fall, when I ascertained with certainty, that, although the land through which it passed was comparatively full of water, the drain actually lost more than it gained in a passage of several chains through it."

So far as authority goes, there seems, with the exception of some advocates of the Keythorpe system, of which an account has been given, to be very little difference of opinion. Mr. Denton says:

"With respect to the direction of drains, I believe very little difference of opinion exists. All the most successful drainers concur in the line of the steepest descent, as essential to effective and economical drainage. Certain exceptions are recognized in the West of England, but I believe it will be found, as practice extends in that quarter, that the exceptions have been allowed in error."

In another place, he says:

"The very general concurrence in the adoption of the line of greatest descent, as the proper course for the minor drains in soils free from rock, would almost lead me to declare this as an incontrovertible principle."

Allusion has been made to cases where we may have to defend ourselves from the flow of water from higher undrained lands of our neighbor. To arrest the flow of mere surface water, an open ditch, or catch-water, is the most effectual, as well as the most obvious mode. There are many instances in New England, where lands upon the lowest slopes of hills are overflowed by water which fell high up upon the hill, and, after passing downward till arrested by rock formation, is borne out again to the surface, in such quantity as to produce, just at the foot of the hill, almost a swamp. This land is usually rich from the wash of the hills, but full of cold water. [153]

To effect perfect drainage of a portion of this land, which we will suppose to be a gentle slope, the first object must be to cut off the flow of water upon or near the surface. An open ditch across the top would most certainly effect this object, and it may be doubtful whether any other drain would be sufficient. This would depend upon the quantity of water flowing down. If the quantity be very great at times, a part of it would be likely to flow across the top of an under-drain, from not having time to percolate downward into it.

In all cases, it is advised, where our work stops upon a slope, to introduce a cross-drain, connecting the tops of all the minor-drains. This cross-drain is called a *header*. The object of it is to cut off the water that may be passing along in the subsoil down the slope, and which would otherwise be likely to pass downward between the system of drains to a considerable distance before finding them. If we suppose the ground saturated with water, and our drains running up the slope and stopping at 4 feet depth, with no header connecting them, they, in effect, stop against 4 feet head of water, and in order to drain the land as far up as they go, must not only take their fair proportion of water which lies between them, but must draw down this 4 feet head beyond them. This they cannot do, because the water from a higher source, with the aid of capillary attraction, and the friction or resistance met with in percolation, will keep up this head of water far above the drained level.

In railway cuttings, and the like, we often see a slope of this kind cut through, without drying the land above the cutting; and if the slope be disposed in alternate layers of sand or gravel, and clay, the water will continue to flow out high up on the perpendicular bank. Even in porous soils of homogeneous character, it will be found that the *head* of water, if we may use the expression, is [154] affected but a short distance by a drain across its flow. Indeed, the whole theory as to the distance of drains apart, rests upon the idea, that the limit to which drains may be expected effectually to operate, is at most but two or three rods.

Whether, in a particular case, a header alone will be sufficient to cut off the flow of water from the higher land, or whether, in addition to the header, an open catch-water may be required, must depend upon the quantity of water likely to flow through or upon the land. An under-drain might be expected to absorb any moderate quantity of what may be termed drainage-water, but it cannot stop a river or mill-stream; and if the earth above the tiles be compact, even water flowing through the soil with rapidity, might pass across it. If there is reason to apprehend this, an open ditch might be added to the header; or, if this is not considered sufficiently scientific or in good taste, a tile-drain of sufficient capacity may be laid, with the ditch above it carefully packed with small stones to the top of the ground. Such a drain would be likely to receive sand and other obstructing substances, as well as a large amount of water, and should, for both reasons, be carried off independently of the small drains, which would thus be left to discharge their legitimate service.

Where it is thought best to connect an open, or surface drain, with a covered drain, it will add much to its security against silt and other obstructions, to interpose a trap or silt-basin at the junction, and thus allow the water to pass off comparatively clean. Where, however, there is a large flow of water into a basin, it will be kept so much in motion as to carry along with it a large amount of earth, and thus endanger the drain below, unless it be very large.

## DISTANCES APART, OR FREQUENCY OF DRAINS. [155]

The reader, who has studied carefully the rival systems of "deep drainage" and "thorough drainage," has seen that the distance of drains apart, is closely connected with that controversy. The greatest variety of opinion is expressed by different writers as to the proper distances, ranging all the way from ten feet apart to seventy, or even more.

Many English writers have ranged themselves on one side or the other of some sharp controversy as to the merits of some peculiar system. Some distinguished geologist has discovered, or thinks he has, some new law of creation by which he can trace the underground currents of water; or some noble noble lord has "patronized" into notice some caprice of an aspiring engineer, and straight-way the kingdom is convulsed with contests to set up or cast down these idols. By careful observation, it is said, we may find "sermons in stones, and good in everything;" and, standing aloof from all exciting controversies, we may often profit, not only by the science and wisdom of our brethren, but also by their errors and excesses. If, by the help of the successes and failures of our English neighbors, we shall succeed in attaining to their present standard of perfection in agriculture, we shall certainly make great advances upon our present position.

As the distances of drains apart, depend manifestly on many circumstances, which may widely vary in the diversity of soil, climate, and cost of labor and materials to be found in the United States, it will be convenient to arrange our remarks on the subject under appropriate heads.

## DISTANCES DEPEND UPON THE NATURE OF THE SOIL.

Water runs readily through sand or gravel. In such soils it easily seeks and finds its level. If it be drawn [156] out at one point, it tends towards that point from all directions. In a free, open sand, you may draw out all the water at one opening, almost as readily as from an open pond.

Yet, even such sands may require draining. A body of sandy soil frequently lies not only upon clay, but in a basin; so that, if the sand were removed, a pond would remain. In such a case, a few deep drains, rightly placed, might be sufficient. This, however, is a case not often met with, though open, sandy soil upon clay is a common formation.

Then there is the other extreme of compact clay, through which water seems scarcely to percolate at all. Yet it has water in it, that may probably soak out by the same process by which it soaked in. Very few soils, of even such as are called clay, are impervious to water, especially in the condition in which they are found in nature. To render them impervious, it is necessary to wet and stir them up, or, as it is termed, *puddle* them. Any soil, so far as it has been weathered — that is, exposed to air, water and frost — is permeable to water to a greater or less degree; so that we may feel confident that the upper stratum of any soil, not constantly under water, will readily allow the water to pass through.

And in considering the "Drainage of Stiff Clays," we shall see that the most obstinate clays are usually so affected by the operation of drainage, that they crack, and so open passages for the water to the drains.

All gravels, black mud of swamps, and loamy soils of any kind, are readily drained.

Occasionally, however — even in tracts of easy drainage, as a whole — deposits are found of some combinations with iron, so firmly cemented together, as to be almost impenetrable with the pickaxe, and apparently impervious [157] to water. Exceptional cases of this nature must be carefully sought for by the drainer.

Whenever a wet spot is observed, seek for the cause, and be satisfied whether it is wet because a spring bursts up from the bottom; or because the subsoil is impervious, and will not allow the surface-water to pass downward. Ascertain carefully the cause of the evil, and then skillfully doctor the disease, and not the symptoms merely. A careful attention to the theory of moisture, will go far to enable us properly to determine the requisite frequency of drains.

## DISTANCES DEPEND UPON THE DEPTH OF THE DRAINS.

The relations of the depth and distance of drains will be more fully considered, in treating of the depth of drains. The idea that depth will compensate for frequency, in all cases, seems now to be abandoned. It is conceded that clay-soils, which readily absorb moisture, and yet are strongly retentive, cannot be drained with sufficient rapidity, or even thoroughness, by drains at any depth, unless they are also within certain distances.

In a porous soil, as a general rule, the deeper the drain, the further it will draw. The tendency of water is to lie level in the soil; but capillary attraction and mechanical obstructions offer constant resistance to this tendency. The farther water has to pass in the soil, the longer time, other things being equal, will be required for the passage. Therefore, although a single deep drain might, in ten days lower the water-line as much as two drains of the same depth, or, in other words, might draw the water all down to its own level, yet, it is quite evident that the two drains might do the work in less time—possibly, in five days. We have seen already the necessity of laying drains deep enough to be below the reach of the subsoil plow and below frost, so that, in the Northern States, the [158] question of shallow drainage seems hardly debatable. Yet, if we adopt the conclusion that four feet is the least allowable depth, where an outfall can be found, there may be the question still, whether, in very open soils, a still greater depth may not be expedient, to be compensated by increased distance.

## DISTANCES DEPEND UPON CLIMATE.

Climate includes the conditions of temperature and moisture, and so, necessarily, the seasons. In the chapter which treats of *Rain*, it will be seen that the quantity of rain which falls in the year is singularly various in different places. Even, in England, "the annual average rain-fall of the wettest place in Cumberland is stated to be 141 inches, while 19½ inches may be taken as the average fall in Essex. In Cumberland, there are 210 days in the year in which rain falls, and in Chiswick, near London, but 124."

A reference to the tables in another place, will show us an infinite variety in the rain-fall at different points of our own country.

If we expect, therefore, to furnish passage for but two feet of water in the year, our drains need not be so numerous as would be necessary to accommodate twice that quantity, unless, indeed, the time for its passage may be different; and this leads us to another point which should ever be kept in mind in New England—the necessity of quick drainage. The more violent storms and showers of our country, as compared with England, have been spoken of when considering *The Size of Tiles*. The sudden transition from Winter to Summer, from the breaking up of deep snows with the heavy falls of rain, to our brief and hasty planting time, requires that our system of drainage should be efficient, not only to take off large quantities of water, but to take them off in a very short [159] time.

How rapidly water may be expected to pass off by drainage, is not made clear by writers on the subject.

"One inch in depth," says an English writer, "is a very heavy fall of rain in a day, and it generally takes two days for the water to drain fully from deep drained land." One inch of water over an acre is calculated to be something more than one hundred tons. This seems, in gross, to be a large amount, but we should expect that an inch, or even two inches of water, spread evenly over a field, would soon disappear from the surface; and if not prevented by some impervious obstruction, it must continue downward.

It is said, on good authority, that, in England, the smallest sized pipes, if the fall be good, will be sufficiently large, at ordinary distances, to carry off all the surplus water. In the author's own fields, where two-inch tiles are laid at four feet depth and fifty feet apart, in an open soil, they seem amply sufficient to relieve the ground of all surplus water from rain, in a very few days. Most of them have never ceased to run every day in the year, but as they are carried up into an undrained plain, they probably convey much more water than falls upon the land in which they lie.

So far as our own observation goes, their flow increases almost as soon as rain begins to fall, and subsides, after it ceases, about as soon as the water in the little river into which they lead, sinks back into its ordinary channel, the freshet in the drains and in the stream being nearly simultaneous. Probably, two-inch pipes, at fifty feet distances, will carry off, with all desirable rapidity, any quantity of water that will ever fall, if the soil be such that the water can pass through it to the distance necessary to find the drains; but it is equally probable that, in a compact clay soil, fifty feet distance is quite too great for sufficiently [160] rapid drainage, because the water cannot get to the drains with sufficient rapidity.

## DISTANCES DEPEND UPON THE COMPARATIVE PRICES OF LABOR AND TILES.

The fact, that the last foot of a four-foot drain costs as much labor as the first three feet, is shown in another chapter, and the deeper we go, the greater the comparative cost of the labor. With tiles at $10 per thousand, the cost of opening and filling a four-foot ditch is, in, round numbers, by the rod, equal to twice the cost of the tiles. In porous soils, therefore, where depth may be made to compensate for greater distance, it is always a matter for careful estimate, whether we shall practice true economy by laying the tiles at great depths, or at the smallest depth at which they will be safe from frost and the subsoil plow, and at shorter distances. The rule is manifest that, where labor is cheap and tiles are dear, it is true economy to dig deep and lay few tiles; and, where tiles are cheap and labor is dear, it is economy to make the number of drains, if possible, compensate for less depth.

## DISTANCES DEPEND UPON SYSTEM.

While we would not lay down an arbitrary arrangement for any farm, except upon a particular examination, and while we would by no means advocate what has been called the gridiron system — of drains everywhere at equal depths and distances — yet some system is absolutely essential, in any operation that approaches to thorough drainage.

If it be only desired to cut off some particular springs, or to assist Nature in some ravine or basin, a deep drain here and there may be expedient; but when any considerable surface is to be drained, there can be no good work without a connected plan of operations. [161]

Mains must be laid from the outfall, through the lowest parts; and into the mains the smaller drains must be conducted, upon such a system as that there may be the proper fall or inclination throughout, and that the whole field shall be embraced.

Again, a perfect *plan* of the completed work, accurately drawn on paper, should always be preserved for future reference. Now it is manifest, that it is impossible to lay out a given field, with proper mains and small drains, dividing the fall as equally as practicable between the different parts of an undulating field, preserving a system throughout, by which, with the aid of a plan, any drain may at any time be traced, without making distances conform somewhat to the system of the whole.

It is easily demonstrable, too, that drains at right angles with the mains, and so parallel with each other, are the shortest possible drains in land that needs uniform drainage. They take each a more uniform share of the water, and serve a greater breadth of soil than when laid at acute angles. While, therefore, it may be supposed that in particular parts of the field, distances somewhat greater or less might be advisable, considered independently, yet in practice, it will be found best, usually, to pay becoming deference to order, "Heaven's first law," and sacrifice something of the individual good, to the leading idea of the general welfare.

In the letter of Mr. Denton, in another chapter, some remarks will be found upon the subject of which we are treating. The same gentleman has, in a published paper, illustrated the impossibility of strict adherence to any arbitrary rule in the distances or arrangement of drains, as follows:

"The wetness of land, which for distinction's sake, I have called 'the water of pressure,' like the water of springs, to which it is nearly allied, can be effectually and cheaply removed only by drains devised for, and [162] devoted to the object. Appropriate deep drains at B B B, for instance, as indicated in the dark vertical lines, are found to do the service of many parallel drains, which as frequently miss, as they hit, those furrows, or 'lips,' in the horizontal out-crop of water-bearing strata which continue to exude wetness after the higher portions are dry.

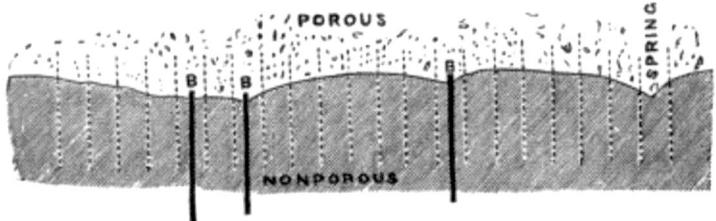

Fig. 35.—The vertical dotted lines show the position of parallel drains.

"A consideration, too, of the varying inclinations of surface, of which instances will frequently occur in the same field, necessitates a departure from uniformity, not in direction only, but in intervals between drains. Take, for instance, the ordinary case of a field, in which a comparatively flat space will intervene between quickly rising ground and the outfall ditch. It is clear that the soak of the hill will pervade the soil of the lower ground, let the system of drainage adopted be what it may; and, therefore, supposing the soil of the hill and flat to be precisely alike, the existence of bottom water in a greater quantity in the lower lands than in the higher, will call for a greater number of drains. It is found, too, that an independent discharge or relief of the water coming from the hill, at B, should always be provided, in order to avoid any impediment by the slower flow of the flatter drains.

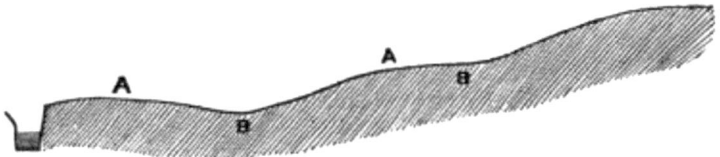

Fig. 36.

"Experience shows that, with few exceptions, hollows, or 'slacks,' observable on the surface, as at B B, have a corresponding undulation of subsoil and that any system which does not provide a direct [163] release for water, which would otherwise collect in and draw towards these spots, is imperfect and unsatisfactory. It is found to be much more safe to depend on relief drains, than on the cutting of drains sufficiently deep through the banks, at A A, to gain a fall at a regular inclination.

"Still, in spite of experience, we often observe a disregard of these facts, even in works which are otherwise well executed to a depth of four feet, but fettered by methodical rules, and I feel compelled to remark, that it has often occurred to me, when I have observed with what diligent examination the rules of depth and distance have been tested, that if more attention had been paid to the *source* of injury, and to the mode of securing an effective and permanent *discharge* of the injurious water, much greater service would be done."

In conclusion, as to distances, we should advise great caution on the part of beginners in laying out their drains. Draining is too expensive a work to be carelessly or unskillfully done. A mistake in locating drains too far apart, brings a failure to accomplish the end in view. A mistake in placing them too near, involves a great loss of labor and money. Consult, then, those whose experience has given them knowledge, and pay to a professional engineer, or some other skillful person, a small amount for aid, which will probably save ten times as much in the end. We have placed our own drains in porous, though very wet soil, at fifty feet distances, which, in most soils, might be considered extremely wide. We are fully satisfied that they would have drained the land as well at sixty feet, except in a few low places, where they could not be sunk four feet for want of fall.

In most New England lands that require drainage, we believe that from 40 to 50 feet distances, with four feet depth, will prove sufficient. Upon stiff clays, we have no experience of our own of any value, although we have a field of the stiffest clay, drained last season at 40 feet distances and four feet depth. In England, this would, probably, prove insufficient, and, perhaps, it will prove so here. [164] One thing is certain, that, at present, there is little land in this country that will pay for drainage by hand labor, at the English distances in clay, of 16 or 20 feet. If our powerful Summer's sun will not somehow compensate in part for distance, we must, upon our clays, await the coming of draining plows and steam.

## DEPTH OF DRAINS.

Cheap and temporary expedients in agriculture are the characteristics of us Americans, who have abundance of land, a whole continent to cultivate, and comparatively few hands and small capital with which to do the work. We erect temporary houses and barns and fences, hoping to find time and means at a future day, to reconstruct them in a more thorough manner. We half cultivate our new lands, because land is cheaper than labor; and it pays best for the present, rather to rob our mother earth, than to give her labor for bread.

The easy and cheap process in draining, is that into which we naturally fall. It is far easier and cheaper to dig shallow than deep drains, and, therefore, we shall not dig deep unless we see good reason to do so. If, however, we carefully study the subject, it will be manifest that superficial drainage is, in general, the result of superficial knowledge of the subject.

Thorough-drainage does not belong to pioneer farming, nor to a cheap and temporary system. It involves capital and labor, and demands skill and system. It cannot be patched up, like a brush fence, to answer the purpose, from year to year, but every tile must be placed where it will best perform its office for a generation. In England, the rule and the habit in all things, is thoroughness and permanency; yet the first and greatest mistake there in drainage was shallowness, and it has required years of [165] experiments, and millions of money, to correct that mistake. If we commit the same folly, as we are very likely to do, we cannot claim even the originality of the blunder, and shall be guilty of the folly of pursuing the crooked paths of their exploration, instead of the straight highway which they have now established. To be sure, the controversy as to the depth of drains has by no means ceased in England, but the question is reduced to this, whether the least depth shall be three feet or four; one party contending that for certain kinds of clay, a three-foot drain is as effectual as a four-foot drain, and that the least effectual depth should be used, because it is the cheapest; while the general opinion of the best scientific and practical men in the kingdom, has settled down upon four feet as the minimum depth, where the fall and other circumstances render it practicable. At the same time, all admit that, in many cases, a greater depth than four feet is required by true economy. It may seem, at first, that a controversy, as to one additional foot in a system of drainage, depends upon a very small point; but a little reflection will show it to be worthy of careful consideration. Without going here into a nice calculation, it may be stated generally as an established fact, that the excavation of a ditch four feet deep, costs twice as much as that of a ditch three feet deep. Although this may not seem credible to one who has not considered the point, yet it will become more probable on examination, and very clear, when the actual digging is attempted. Ditches for tiles are always opened widest at top, with a gradual narrowing to near the bottom, where they should barely admit the tile. Now, the addition of a foot to the depth, is not, as it would perhaps at first appear, merely the addition of the lowest and narrowest foot, but rather of the topmost and widest foot. In other words, a four-foot ditch is precisely a three-foot ditch in size and form, with an [166] additional foot on the top of it, and not a three-foot ditch deepened an additional foot.

The lowest foot of a four-foot ditch is raised one foot higher, to get it upon the surface, than if the ditch were but three feet deep. In clays, and most other soils, the earth grows harder as we go deeper, and this consideration, in practice, will be found important. Again: the small amount of earth from a three-foot ditch, may lie conveniently on one bank near its edge, while the additional mass from a deeper one must be thrown further; and then is to be added the labor of replacing the additional quantity in filling up.

On the whole, the point may be conceded, that the labor of opening and finishing a four-foot drain is double that of a three-foot drain.

Without stopping here to estimate carefully the cost of excavation and the cost of tiles, it may be remarked, that, upon almost any estimate, the cost of labor, even in a three-foot drain in this country, yet far exceeds the cost of tiles: but, if we call them equal, then, if the additional foot of depth costs as much as the first three feet, we have the cost of a four-foot tile-drain fifty per cent. more than that of a three-foot drain. In other words, 200 rods of four-foot drain will cost just as much as 300 rods of three-foot drain. This is, probably, as nearly accurate as any general estimate that can be made at present. The principles upon which the calculations depend, having been thus suggested, it will not be difficult to vary them so as to apply them to the varying prices of labor and tiles, and to the use of the plow or other implements propelled by animals or steam, when applied to drainage in our country.

The earliest experiments in thorough-drainage, in England, were at very small depths, two feet being, for a time, considered very deep, and large tracts were underlaid [167] with tiles at a depth of eighteen, and even twelve inches. It is said, that 10,000 miles of drains, two feet deep and less, were laid in Scotland before it was found that this depth was not sufficient. Of course, the land thus treated was relieved of much water, and experimenters were often much gratified with their success; but it may be safely said now, that there is no advocate known to the public, in England, for a system of drainage of less than three feet depth, and no one advocates a system of drainage of less than four feet deep, except upon some peculiar clays.

The general principle seems well established, that depth will compensate for width; or, in other words, that the deeper the drain, the farther it will draw. This principle, generally correct, is questioned when applied to peculiar clays only. As to them, all that is claimed is, that it is more economical to make the drains but three feet, because they must, even if deep, be near together—nobody doubting, that if four feet deep or more, and near enough, they will drain the land.

In speaking of *clay* soil, it should always be borne in mind, that clay is merely a relative term in agriculture. "A clay in Scotland," says Mr. Pusey, "would be a loam in the South of England." Professor Mapes, of our own country, in the *Working Farmer*, says, "We are convinced, that, with thorough subsoil plowing, no clay soil exists in this country which might not be underdrained to a depth of four feet with advantage."

There can be no doubt, that, with four-foot drains at proper distances, all soils, except some peculiar clays, may be drained, even without reference to the changes produced in the mechanical structure of soil by the operation. There is no doubt, however, that all soils are, by the admission of air, which must always take the place of the water drawn out, and by the percolation of water through them, rendered gradually more porous. Added to this, [168] the subsoil plow, which will be the follower of drainage, will break up the soil to considerable depth, and thus make it more permeable to moisture. But there is still another and more effective aid which Nature affords to the land-drainer, upon what might be otherwise impracticable clays.

This topic deserves a careful and distinct consideration, which it will receive under the title of "Drainage of Stiff Clays."

In discussing the subject of the depth of drains, we are not unmindful of the fact that, in this country, the leaders in the drainage movement, especially Messrs. Delafield, Yeomans, and Johnston, of New York, have achieved their truly striking results, by the use of tiles laid at from two and a half to three feet depth. On the "Premium Farm" of R. J. Swan, of Rose Hill, near Geneva, it is stated that there are sixty-one miles of under-drains, laid from two and a half to three feet deep. That these lands thus drained have been changed in their character, from cold, wet, and unproductive wastes, in many cases, to fertile and productive fields of corn and wheat, sufficiently appears. Indeed, we all know of fields drained only with stone drains two feet deep, that have been reclaimed from wild grasses and rushes into excellent mowing fields. In England and in Scotland, as we have seen, thousands of miles of shallow drains were laid, and were for years quite satisfactory. These facts speak loudly in favor of drainage in general. The fact that shoal drains produce results so striking, is a stumbling-block in the progress of a more thorough system. It may seem like presumption to say to those to whom we are so much indebted for their public spirit, as well as private enterprise, that they have not drained deep enough for the greatest advantage in the end. It would seem that they should know their own farms and their own results better than others. We [169] propose to state, with all fairness, the results of their experiments, and to detract nothing from the credit which is due to the pioneers in a great work.

We cannot, however, against the overwhelming weight of authority, and against the reasons for deeper drainage, which, to us, seem so satisfactory, conclude, that even three feet is, in general, deep enough for under-drains. Three-foot drains will produce striking results on almost any wet lands, but four-foot drains will be more secure and durable, will give wider feeding-grounds to the roots, better filter the percolating water, warm and dry the land earlier in Spring, furnish a larger reservoir for heavy rains, and, indeed, more effectually perform every office of drains.

In reviewing our somewhat minute discussion of this essential point—the proper depth of drains—certain propositions may be laid down with considerable assurance.

## TILES MUST BE LAID BELOW THE REACH OF THE SUBSOIL PLOW.

Let no man imagine that he shall never use the subsoil plow; for so surely as he has become already so much alive to improvement, as to thorough-drain, so surely will he next complete the work thus begun, by subsoiling his land.

The subsoil plow follows in the furrow of another plow, and if the forward plow turn a furrow one foot deep, the subsoil may be run two feet more, making three feet in all. Ordinarily, the subsoil plow is run only to the depth of 18 or 20 inches; but if the intention were to run it no deeper than that, it would be liable to dip much deeper occasionally, as it came suddenly upon the soft places above the drains. The tiles should lie far enough below the deepest path of the subsoil plow, not to be at all disturbed by its pressure in passing over the [170] drains. It is by no means improbable that fields that have already been drained in this country, may be, in the lifetime of their present occupants, plowed and subsoiled by means of steam-power, and stirred to as great a depth as shall be found at all desirable. But, in the present mode of using the subsoil plow on land free from stones, a depth less than three and a half or four feet would hardly be safe for the depth of tile-drains.

## TILES MUST BE LAID BELOW FROST.

This is a point upon which we must decide for our selves. There is no country where drainage is practiced, where the thermometer sinks, as in almost every Winter it does in New England, to 20° below zero (Fahrenheit).

All writers seem to assume that tile-drains must be injured by frost. What the effect of frost upon them is supposed to be, does not seem very clear. If filled with water, and frozen, they must, of course, burst by the expansion of the water in freezing; but it would probably rarely happen, that drainage-water, running in cold weather, could come from other than deep sources, and it must then be considerably above the freezing point. Still; we know that aqueduct pipes do freeze at considerable depths, though supplied from deep springs. Neither these nor gas-pipes are, in our New England towns, safe below frost, unless laid four feet below the surface; and instances occur where they freeze at a much greater depth, usually, however, under the beaten paths of streets, or in exposed positions, where the snow is blown away. In such places, the earth sometimes freezes solid to the depth of even six feet. It will be suggested at once that our fields, and especially our wet lands, do not freeze so deep, and this is true; but it must be borne in mind, that the very reason why our wet lands do not freeze deeper, may [171] be, that they are filled with the very spring-water which makes them cold in Summer, indeed, but is warmer than the air in Winter, and so keeps out the frost. Drained lands will freeze deeper than undrained lands, and the farmer must be vigilant upon this point, or he may have his work ruined in a single Winter.

We are aware, that upon this, as every other point, ascertained facts may seem strangely to conflict. In the town of Lancaster, among the mountains in the coldest part of New Hampshire, many of the houses and barns of the village are supplied with water brought in aqueducts from the hills. We observed that the logs which form the conduit are, in many places, exposed to view on the surface of the ground, sometimes partly covered with earth, but generally very little protected. There has not been a Winter, perhaps in a half century, when the thermometer has not at times been 10° below Zero, and often it is even lower than that. Upon particular inquiry, we ascertained that very little inconvenience is experienced there from the freezing of the pipes. The water is drawn from deep springs in the mountains, and fills the pipes of from one to two-inch bore, passing usually not more than one or two hundred rods before it is discharged, and its warmth is sufficient, with the help of its usual snow covering, to protect it from the frost.

We have upon our own premises an aqueduct, which supplies a cattle-yard, which has never been covered more than two feet deep, and has never frozen in the nine years of its use. We should not, therefore, apprehend much danger from the freezing of pipes, even at shallow depths, if they carry all the Winter a considerable stream of spring-water; but in pipes which take merely the surface water that passes into them by percolation, we should expect little or no aid from the water in preventing frost. The water filtering downward in Winter must be nearly [172] at the freezing point; and the pipes may be filled with solid ice, by the freezing of a very small quantity as it enters them.

Neither hard-burnt bricks nor hard-burnt tiles will crumble by mere exposure to the Winter weather above ground, though soft bricks or tiles will scarcely endure a single hard frost. Too much stress cannot be laid upon the importance of using hard-burnt tiles only, as the failure of a single tile may work extensive mischief.

Writers seem to assume, that the freezing of the ground about the drains will displace the tiles, and so destroy their continuity, and this may be so; though we find no evidence, perhaps, that at three or four feet, there is any disturbance of the soil by freezing. We dig into clay, or into our strong subsoils, and find the earth, at three feet deep, as solid and undisturbed as at twice that depth, and no indication that the frost has touched it, though it has felt the grip of his icy fingers every year since the Flood. With these suggestions for warning and for encouragement, the subject must be left to the sound judgment of the farmer or engineer upon each farm, to make the matter so safe, that the owner need not have an anxious thought, as he wakes in a howling Winter night, lest his drains should be freezing.

Finally, in view of the various considerations that have been, suggested, as well as of the almost uniform authority of the ablest writers and practical men, it is safe to conclude, that, in general, in this country, wherever sufficient outfall can be had, *four feet above the top of the tiles should be the minimum depth of drains.*

# CHAPTER VIII [173]
# ARRANGEMENT OF DRAINS.

Necessity of System.—What Fall is Necessary.—American Examples.—Outlets.—Wells and Relief-Pipes.—Peep holes.—How to secure Outlets.—Gate to Exclude Back-Water.—Gratings and Screens to keep out Frogs, Snakes, Moles, &c.—Mains, Submains, and Minors, how placed.—Capacity of Pipes.—Mains of Two Tiles.—Junction of Drains.—Effect of Curves and Angles on Currents.—Branch Pipes.—Draining into Wells or Swallow Holes.—Letter from Mr. Denton.

As every act is, or should be, a part of a great plan of life, so every stake that is set, and every line laid in the field, should have relation not only to general principles, but also to some comprehensive plan of operations.

Assuming, then, that the principles advocated in this treatise are adopted as to the details, that the depth preferred is not less than four feet—that the direction preferred is up and down the slope—that the distance apart may range from fifteen to sixty feet, and more in some cases, according to the depth of drains and the nature of the soil—that no tiles smaller than one and a half inch bore will be used, and none less than two inches except for the first one hundred yards, there still remains the application of all these principles to the particular work in hand. With the hope of assisting the deliberations of the farmer on this point, some additional suggestions will be made under appropriate heads.

## ARRANGEMENT MUST HAVE REFERENCE TO SYSTEM.

The absolute necessity of some regularity of plan in our work, must be manifest. Without system, we can never, [174] in the outset, estimate the cost of our operation; we can never proportion our tiles to the quantity of water that will pass through them; we can never find the drains afterwards, or form a correct opinion of the cause of any failure that may await us.

We prefer, in general, where practicable, parallel lines for our minor drains, at right angles with the mains, because this is the simplest and most systematic arrangement; but the natural ravines or water-courses in fields, seldom run parallel with each other, or at right angles with the slope of the hills, so that regular work like this, can rarely be accomplished.

If the earth were constructed of regular slopes, or plains of uniform character, we could easily apply to it all our rules; but, broken as it is into hills and valleys, filled with stones here, with a bank of clay there, and a sand-pit close by, we are obliged to sacrifice to general convenience, often, some special abstract rule.

We prefer to run drains up and down the slope; but if the field be filled with undulations, or hills with various slopes, we may often find it expedient, for the sake of system, to vary this course.

If the question were only as to one single drain, we could adjust it so as to conform to our perfect ideal; but as each drain is, as it were, an artery in a complicated system, which must run through and affect every part of it, all must be located with reference to every other, and to the general effect.

Keeping in mind, then, the importance of some regular system that shall include the whole field of operation, the work should be laid out, with as near a conformity to established principles as circumstances will permit.

## ARRANGEMENT MUST HAVE REFERENCE TO THE FALL.

In considering what fall is necessary, and what is desirable, we have seen, that although a very slight inclination [175] may carry off water, yet a proportionably larger drain is necessary as the fall decreases, because the water runs slower.

"It is surprising," says Stephens, "what a small descent is required for the flow of water in a well-constructed duct. People frequently complain that they cannot find sufficient fall to carry off the water from the drains. There are few situations where a sufficient fall cannot be found if due pains are exercised. It has been found in practice, that a water-course thirty feet wide and six feet deep, giving a transverse sectional area of one hundred and eighty square feet, will discharge three hundred cubic yards of water per minute, and will flow at the rate of one mile per hour, with a fall of no more than *six inches per mile*."

Messrs. Shedd and Edson, of Boston, have superintended some drainage works in Milton, Mass., where, after obtaining permission to drain through the land of an adjacent owner, not interested in the operation, they could obtain but three inches fall in one hundred feet, or a half inch to the rod, for three quarters of a mile, and this only by blasting the ledges at the outlet. This fall, however, proves sufficient for perfect drainage, and by their skill, a very unhealthful swamp has been rendered fit for gardens and building-lots. In another instance, in Dorchester, Mass., Mr. Shedd informs us that in one thousand feet, they could obtain only a fall of two inches for their main, and this, by nice adjustment, he expects to make sufficient. In another instance, he has found a fall of two and a half inches in one hundred feet, in an open paved drain to be effectual.

It is certainly advisable always to divide the fall as even as possible throughout the drains, yet this will be found a difficult rule to follow. Very often we have a space of nearly level ground to pass through to our outfall; and, usually, the mains, in order that the minor drains may be carried into them from both sides, must follow up the natural valleys in the field, thus controlling, in a great measure, [176] our choice as to the fall. We are, in fact, often compelled to use the natural fall nearly as we find it.

It is thought advisable to have the mains from three to six inches lower than the drains discharging into them, so that there may be no obstruction in the minor drains by the backing up of water, and the consequent deposition of sand or other obstructing substances. Wherever one stream flows into another, there must be more or less interruption of the course of each. If the water from the minors enters the main with a quick fall, the danger of obstruction in the minor, at least, is much lessened. A frequent cause of partial failure of drains, is their not having been laid with a regular inclination. If, instead of a gradual and uniform fall, there should be a slight rising in the bed of a drain, the descending water will be interrupted there till it accumulate so high as to be above the level of the rising. At this point, therefore, the water must have a tendency to press out of the drains, and will deposit whatever particles of sand or other earthy matter it may bring down.

Drains must, therefore, be so arranged, that in cutting them, their beds may be as nearly as possible, straight, or, at least, have a constant, if not a regular and equal fall.

## ARRANGEMENT MUST HAVE REFERENCE TO THE OUTLET.

All agree that it is best to have but few general outlets. "In the whole process of draining," says an engineer of experience, "there is nothing so desirable as permanent and substantial work at the point of discharge." The outlet is the place, of all others, where obstruction is most likely to occur. Everywhere else the work is protected by the earth above it, but here it is exposed to the action of frost, to cattle, to mischievous boys, to reptiles, as well as to the obstructing deposits which are discharged from the drains themselves. In regular work, under the direction [177] of engineers, iron pipes, with swing gratings set in masonry, are used, to protect permanently this important part of the system of drainage.

It may often be convenient to run parallel drains down a slope, bringing each out into an open ditch, or at the bottom of some bank, thus making a separate outlet for each. This practice, however, is strongly deprecated. These numerous outlets cannot be well protected without great cost; they will be forgotten, or, at least, neglected, and the work will fail.

Regarding this point, of few and well-secured outlets, as of great importance, the arrangement of all the drains must have reference to it. When drains are brought down a slope, as just suggested, let them, instead of discharging separately, be crossed, near the foot of the slope, by a sub-main running a little diagonally so as to secure sufficient fall, and so carried into a main, or discharged at a single outlet.

It may be objected, that by thus uniting the whole system, and discharging the water at one point, there may be difficulty in ascertaining by inspection, whether any of the drains are obstructed, or whether all are performing their appropriate work. There is prudence and good sense in this suggestion, and the objection may be obviated by placing *wells*, or "peep-holes," at proper intervals, in which the flow of the water at various points may be observed. On the subject of wells and peep-holes, the reader will find in another chapter a more particular description of their construction and usefulness.

The position of the outlet must, evidently, be at a point sufficiently low to receive all the water of the field; or, in other words, it must be the lowest point of the work. It will be fortunate, too, if the outlet can be at the same time high enough to be at all times above the back-water of the stream, or pond, or marsh, into which it [178] empties; and high enough, too, to be protected by solid earth about it. In any case, great care should be taken to make the outlet secure and permanent. The process of thorough-drainage is expensive, and will only repay cost, upon the idea that it is permanent—that once well done, it is done forever. The tiles may be expected to operate well, for a lifetime; and the outlet, the only exposed portion of the work, should be constructed to endure as long as the rest.

It is true that this portion of the work may be reached and repaired more conveniently than the tiles themselves; but it must be remembered that the decay of the outlet obstructs the flow of the water, produces a general stagnation throughout the drains, and so may cause their permanent obstruction at various points, hard to be ascertained, and difficult to be reached. Considering our liability to neglect such things as perish by a gradual decay, as well as the many accidental injuries to which the outlet is exposed, there is no security but in a solid and permanent structure at the first.

To illustrate the importance attached to this point in England, as well as to indicate the best mode of securing the outlet, the drawings below have been taken from a pamphlet by Mr. Denton. Fig. 37 represents the mode of constructing the common small outlets of field drainage.

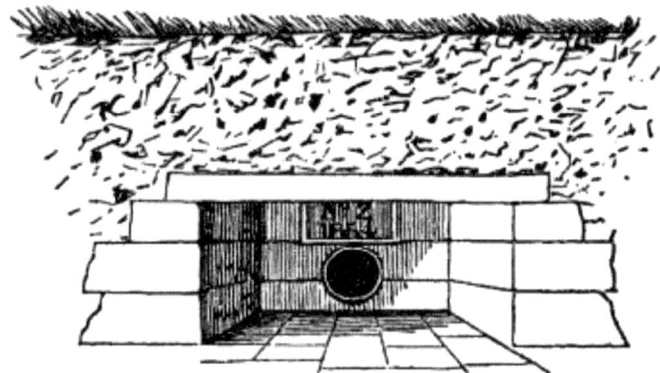

Fig. 37.—Small Outlet.

The distinguished engineer, of whose labors we have so [179] freely availed ourselves, remarks as follows upon the subject:

"Too many outlets are objectionable, on account of the labor of their maintenance: too few are objectionable, because they can only exist where there are mains of excessive length. A limit of twenty acres to an outlet, resulting in an average of, perhaps, fourteen acres, will appear, by the practices of the best drainers, to be about the proper thing. If a shilling an acre is reserved for fixing the outlets, which should be *iron pipes, with swing gratings*, in masonry, very substantial work may be done."

Figures 38 and 39 represent the elevation and section of larger outlets, used in more extensive works.

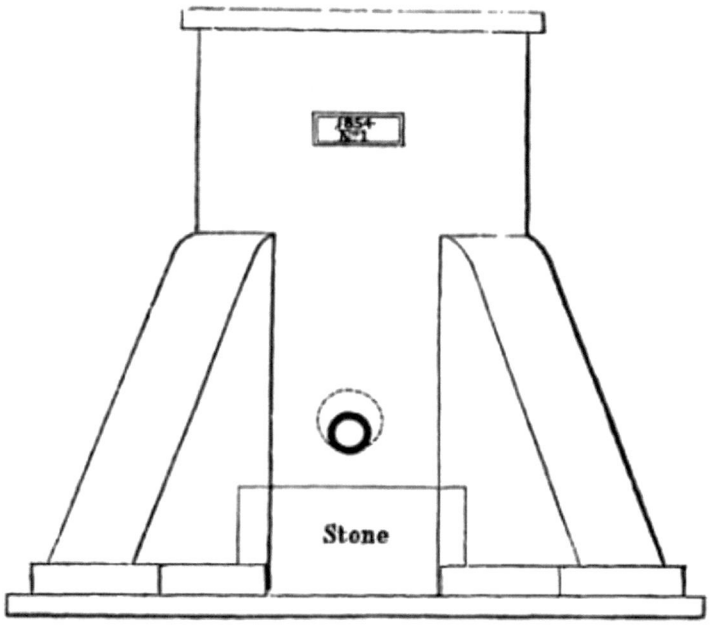

Fig. 38. — Large Outlet.

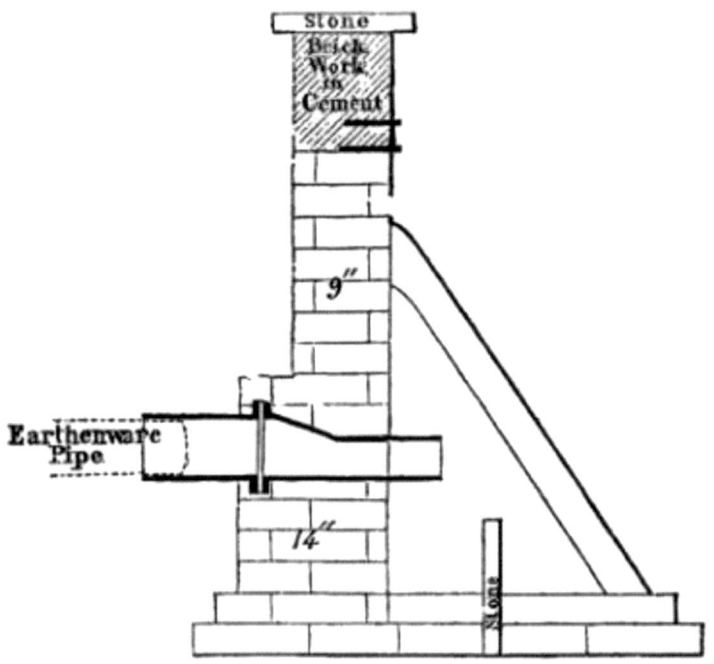

Fig. 39.—Large Outlet.

It is almost essential to the efficiency of drains, that there be fall enough beyond the outlet to allow of the quick flow of the water discharged. At the outlet, must be deposited whatever earth is brought down by the drains; and, in many cases, the outlet must be at a swamp or pond. If no decided fall can be obtained at the outlet, [180] there must be care to provide and keep an open ditch or passage, so that the drainage-water may not be dammed back in the drains. It is advised, even, to follow down the bank of a stream or river, so as to obtain sufficient fall, rather than to have the outlet flooded, or *back-water* in the drains. Still, there may be cases where it will be impossible to have an outlet that shall be always above the level of the river or pond which may receive the drainage water. If the outlet must be so situated as to be at times overflowed, great care should be taken to excavate a place at the outlet, into which any deposits brought down by the drain, may fall. If the outlet be level with the ground beyond it, the smallest quantity of earth will

190

operate as a dam to keep back the water. Therefore, at the outlet, in such cases, a small well of brick or stonework should be constructed, into which the water should pour. There, even if the water stand above the outlet, [181] will be deposited the earth brought along in the drain. This well must at times, when the water is low, be cleared of its contents, and kept ready for its work.

The effect of back-water in drains cannot ordinarily be injurious, except as it raises the water higher in the land, and occasions deposits of earthy matter, and so obstructs the drains. We have in mind now, the common case of water temporarily raised, by Winter flowage or by Summer freshets.

It should be remembered that even when the outlet is under water, if there is any current in the stream into which the drain empties, there must be some current in the drain also; and even if the drain discharge into a still pond, there must be a current greater or less, as water from a level higher than the surface of the pond, presses into the drains. Generally, then, under the most unfavorable circumstances, we may expect to have some flow of water through the pipes, and rarely an utter stagnation. If, then, the tiles be carefully laid, so as to admit only well-filtered water, there can be but little deposit in the drain; and a temporary stagnation, even, will not injure them, and a trifling flow will keep them clean. Much will depend, as to the obstruction of drains, in this, and indeed in all cases, upon the internal smoothness, and upon the nice adjustment of the pipes. In case of the drainage of marshes, and other lands subject to sudden flood, a flap, or gate, is used to exclude the water of flowage, until counterbalanced by the drainage-water in the pipes.

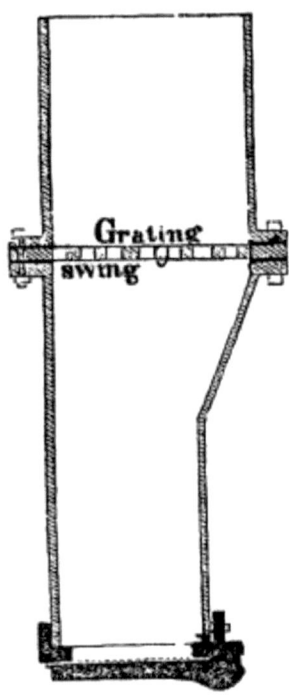

Fig. 40.—Outlet Pipe with Flap to Exclude Flood-water.

We are quite sure that it is not in us a work of supererogation to urge upon our farmers the importance of careful [182] attention to this matter of outlets. This is one of that class of things which will never be attended to, if left to be daily watched. We Americans have so much work to do, that we have no time to be careful and watchful. If a child fall into the fire, we take time to snatch him out. If a sheep or ox get mired in a ditch, we leave our other business, and fly to the rescue. Even if the cows break into the corn, all hands of us, men and boys and dogs, leave hoeing or haying, and drive them out. And, by the way, the frequency with which most of us have had occasion to leave important labors to drive back unruly cattle, rendered lawless by neglect of our fences, well illustrates a national characteristic. We are earnest, industrious, and intent on *doing*. We can look forward to accomplish any labor, however difficult, but lack the conservatism which preserves the fruit of our labors—the "old fogyism" which puts on its spectacles with most careful adjustment, after wiping the glasses for a clear sight, and at stated periods, revises its affairs to see if some screw has not worked loose. A steward on a large estate, or a corporation agent, paid for inspecting and superintending, may be relied upon to examine his drainage works, and maintain them in repair; but no farmer in this country, who labors with his own hands, has time even for this most essential duty. His policy is, to do his work now, while he is intent upon it, and not trust to future watchfulness.

We speak from personal experience in this matter of outfalls. Our first drains ran down into a swamp, and the fall was so slight, that the mains were laid as low as possible, so that at every freshet they are overflowed. We have many times, each season, been compelled to go down, with spade and hoe, and clear away the mud which has been trodden up by cattle around the outlet. Although a small river flows through the pasture, the cows find [183] amusement, or better water, about these drains, and keep us in constant apprehension of a total obstruction of our works. We propose to relieve ourself of this care, by connecting the drains together, and building one or more reliable outlets.

## GRATINGS OR SCREENS AT THE OUTLET.

There are many species of "vermin," both "creeping things" and "slimy things, that crawl with legs," which seem to imagine that drains are constructed for their especial accommodations. In dry times, it is a favorite amusement of moles and mice and snakes, to explore the devious passages thus fitted up for them, and entering at the capacious open front door, they never suspect that the spacious corridors lead to no apartments, that their accommodations, as they progress, grow "fine by degrees and beautifully less," and that these are houses with no back doors, or even convenient places for turning about for a retreat. Unlike the road to Hades, the descent to which is easy, here the ascent is inviting; though, alike in both cases, "*revocare gradum, hoc opus hic labor est.*" They persevere upward and onward till they come, in more senses than one, to "an untimely end." Perhaps stuck fast in a small pipe tile, they die a nightmare death; or, perhaps overtaken by a shower, of the effect of which, in their ignorance of the scientific principles of drainage, they had no conception, they are drowned before they have time for deliverance from the straight in which they find themselves, and so are left, as the poet strikingly expresses it, "to lie in cold *obstruction* and to rot."

In cold weather, water from the drains is warmer than the open ditch, and the poor frogs, reluctant to submit to the law of Nature which requires them to seek refuge in mud and oblivious sleep, in Winter, gather round the outfalls, as they do about springs, to bask in the warmth [184] of the running water. If the flow is small, they leap up into the pipe, and follow its course upward. In Summer, the drains furnish for them a cool and shady retreat from the mid-day sun, and they may be seen in single file by scores, at the approach of an intruding footstep, scrambling up the pipe. Dying in this way, affects these creatures, as "sighing and grief" did Falstaff, "blows them up like a bladder;" and, like Sampson, they do more mischief in their death, than in all their life together. They swell up, and stop the water entirely, or partially dam it, so that the effect of the work is impaired.

To prevent injuries from this source, there should be, at every outlet, a grating or screen of cast iron, or of copper wire, to prevent the intrusion of vermin. The screen should be movable, so that any accumulation in the pipe may be removed. An arrangement of this kind is shown in Fig. 40, as used in England. We know of nothing of the kind used in this country. For ourself, we have made of coarse wire-netting, a screen, which is attached to the pipe by hinges of wire. Holes may be bored with a bit through even a hard tile, or a No. 9 wire may be twisted firmly round the end of it, and the screen thus secured.

This has thus far, been our own poor and unsatisfactory mode of protecting our drains. It is only better than none, but it is not permanent, and we hope to see some successful invention that may supply this want. So far as we have observed, no such precaution is used in this country; and in England, farmers and others who take charge of their own drainage works, often run their pipes into the mud in an open ditch, and trust the water to force its own passage.

## OF WELLS AND RELIEF PIPES.

In draining large tracts of land of uniform surface, it is often convenient to have single mains, or even minors, [185] of great length. Obstructions are liable to occur from various causes: and, moreover, there is great satisfaction in being certain that all is going right, and in watching the operation of our subterranean works. It is a common practice, and to be commended, to so construct our drains, that they may be inspected at suspicious points, and that so we may know their real condition.

For this purpose, wells, or traps, are introduced at suitable points, into which the drains discharge, and from which the water proceeds again along its course.

These are made of iron, or of stone or brick work, of any size that may be thought convenient, secured by covers that may be removed at pleasure.

Where there is danger of obstruction below the wells, relief pipes may be introduced, or the wells may overflow, and so discharge temporarily, the drainage water. These wells, sometimes called silt basins, or traps, are frequently used in road drainage, or in sewers where large deposits are made by the drainage water. The sediment is carried along and deposited in the traps, while the water flows past.

These traps are large enough for a man to enter, and are occasionally cleared of their contents.

When good stone, or common brick, are at hand, occasional wells may be easily constructed. Plank or timber might be used; and we have even seen an oil cask made to serve the purpose temporarily. In most parts of New England, solid iron castings would not be expensive.

The water of thorough-drainage is usually as pure as spring-water, and such wells may often be conveniently used as places for procuring water for both man and beast, a consideration well worth a place in arrangements so permanent as those for drainage.

The following figures represent very perfect arrangements of this kind, in actual use. [186]

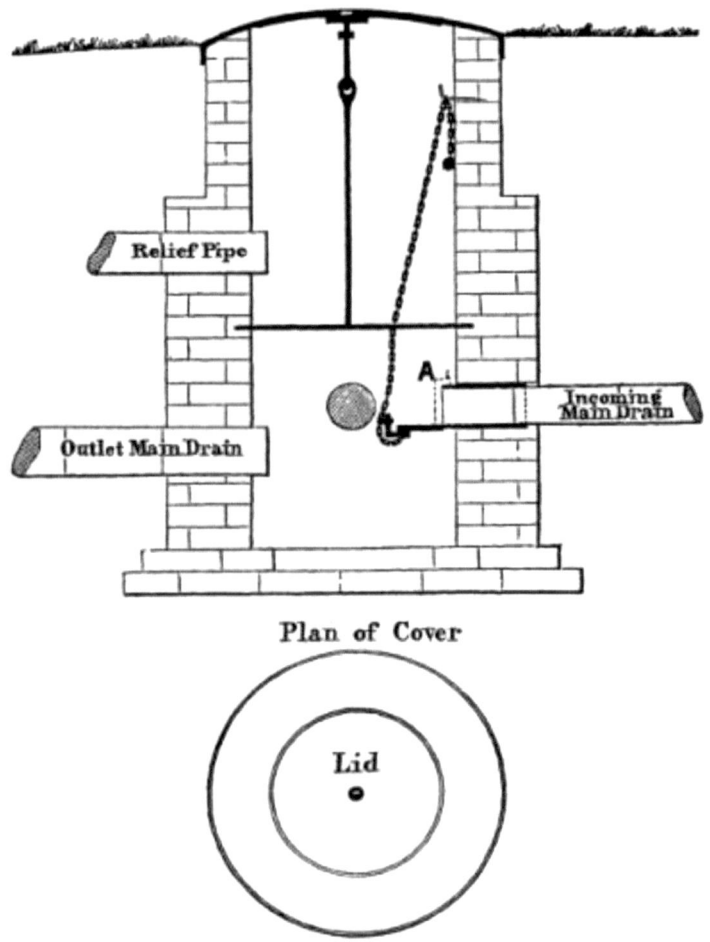

**Plan of Cover**

Figs. 41 & 42.—Well with Silt Basin, or Trap, and Cover.

The flap attached to a chain at A, is designed to close the incoming drain, so as to keep back the water, and thus flush the drain, as it is termed, by filling it with water, and then suddenly releasing it. It is found that by this process, obstructions by sand, and by peroxide of iron, may be brought down from the drains, when the flow is usually feeble. [187]

## SMALL WELLS, OR PEEP-HOLES.

By the significant, though not very elegant name of peep-holes, are meant openings at junctions, or other convenient points, for watching the pulsations of our subterranean arteries.

In addition to the large structures of wells and traps, such as have been represented, we need small and cheap arrangements, by which we may satisfy ourselves and our questioning friends and neighbors, that every part of our buried treasure, is steadily earning its usury. It is really gratifying to be able to allow those who "don't see how water can get into the tiles," and who inquire so distrustfully whether you "don't think that land on the hill would be just as dry without the drains," to satisfy themselves, by actually seeing, that there is a liberal flow through all the pipes, even in the now dry soil. And then, again,

> "The best laid schemes o' mice an' men
> Gang aft agley."

and drains will get obstructed, by one or other of the various means suggested in another place. It is then convenient to be able to ascertain with certainty, and at once, the locality of the difficulty, and this may be done by means of peep-holes.

These may be formed of cast iron, or of well-burnt clay, or what is called stone-ware, of 4, 6, or 10 inches internal diameter, and long enough to reach from the bottom of the drain to the surface, or a little above it.

The drain or drains, coming into this little well, should enter a few inches above the pipe which carries off the water, so that the incoming stream may be plainly seen. A strong cover should be fitted to the top, and secured so as not to cause injury to cattle at work or feeding on the land. The arrangement will be at once seen by a sketch given on the following page. [188]

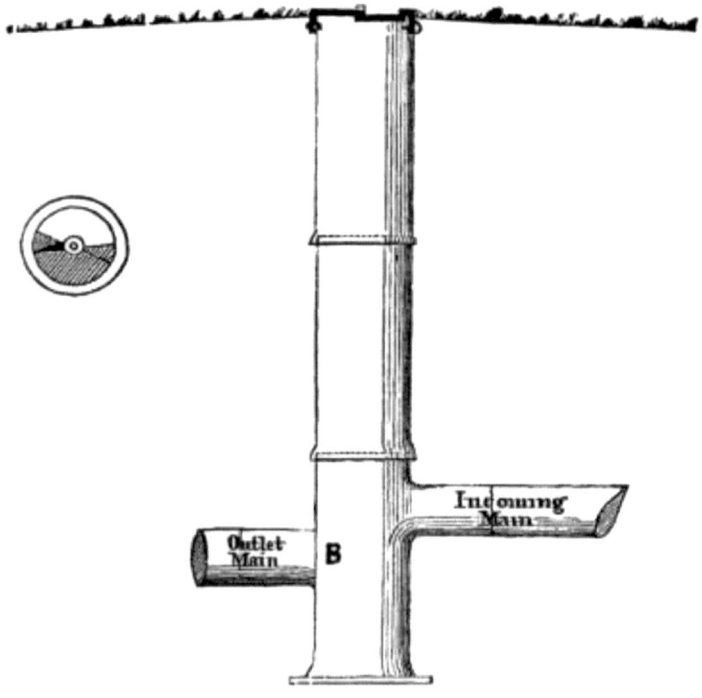

Figs. 43 & 44.—Small Well, or Peep-hole, and Cover.

In our own fields, we have adopted several expedients to attain this object of convenient inspection. In one case, where we have a sub-main, which receives the small drains of an acre of orchard, laid at nearly five feet depth, we sunk two 40-gallon oil casks, one upon the other, at the junction of this sub-main with another, and fitted upon the top a strong wooden cover. The objections to this contrivance are, that it is temporary; that it occupies too much room; and that it is more expensive than a well of cast iron or stone-ware of proper size.

In another part of the same field, we had a spring of excellent water, where, "from the time whereof the memory of man runneth not to the contrary," people had fancied they found better water to drink, than anywhere else. It is near a ravine, through which a main

[189] drain is located, and which is now graded up into convenient plow land.

To preserve this spring for use in the Summer time, we procured a tin-worker to make a well, of galvanized iron, five feet long and ten inches diameter, into which are conducted the drain and the spring. A friendly hand has sketched it for us very accurately; thus:

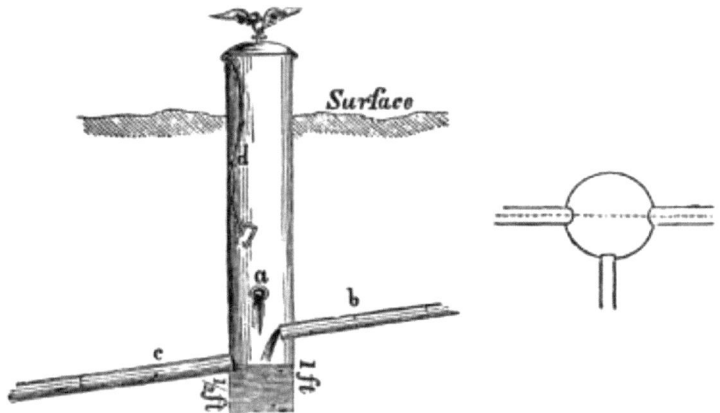

Figs. 45 & 46. — How to Preserve a Spring in a Drained Field.

The spring is brought in at *a* by a few tiles laid into the bank where the water naturally bursts out. The pipe *b* brings in the drain, which always flows largely, and the pipe *c* carries away the water. The small dipper, marked *d*, hangs inside the well, and is used by every man, woman, and boy, who passes that way. The spring enters six inches above the drain, for convenience in catching its water to drink.

By careful observation the present Winter of 1858-9, the impression that there is some peculiar quality in this water is confirmed, for it is ascertained that it is six degrees warmer in cold weather than any other water upon the farm. The spring preserves a temperature of [190] about 47°, while the drain running through the same well, and the other drains in the field, and the well at the house, vary from 39° to 42°.

We confess to the weakness of taking great satisfaction in sipping this water, cool in Summer and warm in Winter, and in watching the mingled streams of spring and drainage water, and listening as we pass by, to their tinkling sound, which, like the faithful watchman of the night, proclaims that "all is well."

## POSITION AND SIZE OF THE MAINS.

Having fixed on the proper position of the outlet, for the whole, or any portion of our work, the next consideration is the location of the drains that shall discharge at that point. It is convenient to speak of the different drains as *mains*, *sub-mains*, and *minors*. By *mains*, are understood the principal drains, of whatever material, the office of which is, to receive and carry away water collected by other drains from the soil. By *minors*, are intended the small drains which receive the surplus water directly from the soil. By *sub-mains*, are meant such intermediate drains as are frequently in large fields, interposed across the line of the minors, to receive their discharge, and conduct their water to the mains.

They are principally used, where there is a greater length of small drains in one direction than it is thought expedient to use; or where, from the unequal surface, it is necessary to lay out subordinate systems of drains, to reach particular localities.

Whether after the outlet is located, the mains or minors should next be laid out, is not perhaps very important. The natural course would seem to be, to lay out the mains according to the surface formation of the land, through the principal hollows of the field, although we have high authority for commencing with the minors, and allowing [191] their appropriate direction to determine the location of the mains.

This is, however, rather a question of precedence and etiquette, than of practical importance. The only safe mode of executing so important a work as drainage, is by careful surveys by persons of sufficient skill, to lay out the whole field of operations, before the ground is broken; to take all the levels; to compare all the different slopes; consider all the circumstances, and arrange the work as a systematic whole. Generally, there will be no conflict of circumstances, as to where the mains shall be located. They must be lower than the minors, because they receive their water. They must ordinarily run across the direction of the minors, either at right angles or diagonally, because otherwise they cannot receive their discharge. If, then, in general, the minors, as we assume, run down the slope, the mains must run at the foot of the slope and across it.

It will be found in practice, that all the circumstances alluded to, will combine to locate the mains across the foot of regular slopes; and whether in straight or curved lines, along through the natural valleys of the field.

In locating the mains, regard must always be had to the quantity of water and to the fall. Where a field is of regular slope, and the descent very slight, it will be necessary, in order to gain for the main the requisite fall, to run it diagonally across the bottom of the slope, thus taking into it a portion of the fall of the slope. If the fall requires to be still more increased, often the main may be deepened towards the outlet, so as to gain fall sufficient, even on level ground.

If the fall is very slight, the size of the main may be made to compensate in part for want of fall, for it will not be forgotten, that the capacity of a pipe to convey water depends much on the velocity of the current, and [192] the velocity increases in proportion to the fall. If the fall and consequent velocity be small, the water will require a larger drain to carry it freely along. The size of the mains should be sufficient to convey, with such fall as is attainable, the greatest quantity of water that may ever be expected to reach them. Beyond this, an increase of size is rather a disadvantage than otherwise, because a small flow of water runs with more velocity when compressed in a narrow channel, than when broadly spread, and so has more power to force its way, and carry before it obstructing substances.

We have seen, in considering the size of tiles, that in laying the minor drains, their capacity to carry all the water that may reach them is not the only limit of their size. A one-inch tile might in many cases be sufficient to conduct the water; but the best drainers, after much controversy on the point, now all agree that this is a size too small for prudent use, because so small an opening is liable to be obstructed by a very slight deposit from the water, or by a slight displacement, and because the joints furnish small space for the admission of water.

Mains, however, being designed merely to carry off such water as they may receive from other drains, may in general be limited to the size sufficient to convey such water, at the greatest flow. It might seem a natural course, to proportion the capacity of the main to the capacity of the smaller drains that fall into it; and this would be the true rule, were the small drains expected to run full.

If our smallest drain, however, be of two-inch, or even one and a half inch bore, it can hardly be expected to fill at any time, unless of great length, or in some peculiarly wet place. Considering, then, what quantity of water will be likely to be conducted into the main, proportion the main not to the capacity of all the smaller drains [193] leading into it, but to the probable maximum flow—not to what they *might* bring into it, but to what they *will* bring.

If the mains be of three-inch pipes, other things being equal, their capacity is nine times that of a one-inch pipe, and two and a quarter times the capacity of a two-inch pipe.

A three-inch main may, then, with equal fall and directness, be safely relied on to carry nine streams of water equal each to one inch diameter, or two and a quarter streams, equal to a two-inch stream. The three-inch main will, in fact, from the less amount of friction, carry much more than this proportion.

The allowance to be made for a less fall in the mains, has already been adverted to, and must not be overlooked. It is believed that the capacity of a three or four-inch pipe to convey water, is in general likely to be much under-estimated.

It is a common error, to imagine that some large stone water-course must be necessary to carry off so large a flow as will be collected by a system over a ten or twenty-acre field. Any one, however, who has watched the full flow of even a three-inch pipe, and observed the water after it has fallen into a nearly level ditch, will be aware, that what seems in the ditch a large stream, impeded as it is by a rough, uneven bottom, may pass through a three inch opening of smooth, well-jointed pipes. When we consider that a four-inch pipe is four times as capacious as a two-inch pipe, and sixteen times as large as a one-inch pipe, we may see that we may accommodate any quantity of water that may be likely anywhere to be collected by drainage, without recourse to other materials than tiles.

When one three or four-inch pipe is not sufficient to convey the water, mains may conveniently be formed of [194] two or more tiles of any form. A main drain is sometimes formed by combining two horse-shoe tiles, with a tile sole or slate between them, to prevent slipping, as in fig. 47.

Fig. 47.

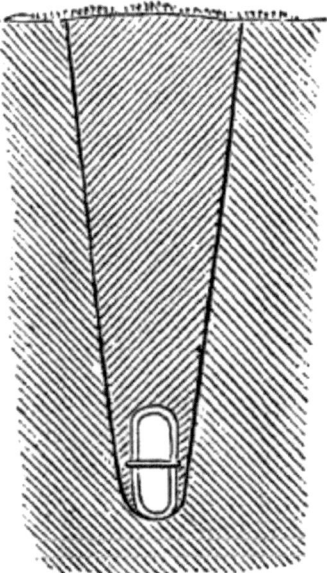

Fig. 48.

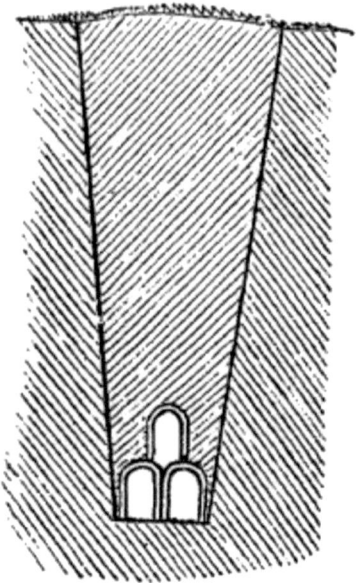

Main Drain of two or more Horse-shoe Tiles.

The combinations represented in the above figures, will furnish sufficient suggestions to enable any one to select or arrange such forms as may be deemed best suited to the case in hand. Where the largest obtainable tile is not large enough, two or more lines of pipes may be laid abreast.

## POSITION OF THE MINOR DRAINS.

Assuming that it is desirable to run the small drains, as far as practicable, up and down the slope, the following directions, from the Cyclopedia of Agriculture, are given:

"There is a very simple mode of laying out these (the minor drains), which will apply to most cases, or, indeed, to all, although in some its application may be more difficult. The surface of each field must be regarded as being made up of one or more planes, as the case may be, for each of which the drains should be laid out separately. Level lines are to be set out, a little below the upper edge of each of these planes, and the drains must be then made to cross these lines at right angles. By this means, the drains will run in the line of the greatest slope, no matter how distorted the surface of the field may be."

Much is said, in the English books, about "furrows," [195] and the "direction of the furrows," in connection with the laying out of drains. Much of the land in England, especially in moist places, was formerly laid up by repeated plowings, into ridges varying in breadth from ten to twenty feet, so as to throw off, readily, the water from the surface.

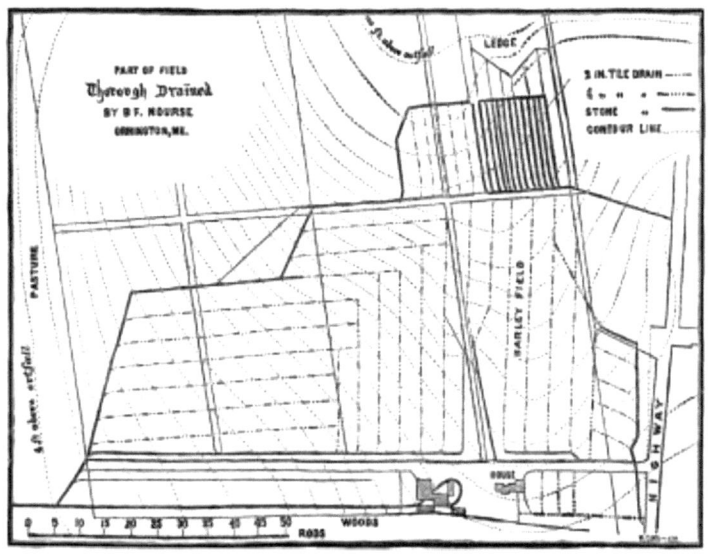

These ridges were sometimes so high, that two boys in opposite furrows, between the ridges, could not see each other. In draining lands thus ridged, it is found far more easy to cut the ditches in the furrows, rather than across or upon the ridges. After thorough-drainage, in most localities, these ridges and furrows are dispensed with. The fact is, probably, only important here, as explaining the constant reference by English writers to this mode of working the land.

Whether we shall drain "down the furrows," or "across the ridges," is not likely to be inquired of, by Americans.

The accompanying diagram represents a field of about thirty acres, as drained by the owner, B. F. Nourse, Esq., of Orrington, Me., a particular description of which will be found in another place.

The curves of the ends of the minors, at their junction with the mains, will indicate their course—the minors curving always so as to more nearly coincide, in course, with the current of water in the mains.

## THE JUNCTION OF DRAINS.

Much difficulty arises in practice, as to connecting, in a secure and satisfactory manner, the smaller with the larger drains. It has already been suggested, that the streams should not meet at right angles, but that a bend should be made in the smaller drain, a few feet before it enters the main, so as to introduce the water of the small drain in the direction of the current in the main. In another place, an instance is given where it was found that a [196] quantity of water was discharged with a turn, or junction with a gentle curve, in 100 seconds, that required 140 seconds with a turn at right angles; and that while running direct, that is, without any turn, it was discharged in 90 seconds. This is given as a mere illustration of the principle, which is obvious enough. Different experiments would vary with the velocity, quantity of water, and smoothness of the pipe; but nothing is more certain, than that every change of direction impedes velocity.

Thus we see that if we had but a single drain, the necessary turns should be curved, to afford the least obstruction.

Where the drain enters into another current, there is yet a further obstruction, by the meeting of the two streams. Two equal streams, of similar velocity and size, thus meeting at right angles, would have a tendency to move off diagonally, if not confined by the pipe; and, confined as they are, must both be materially retarded in their flow. In whatever manner united, there must be much obstruction, if the main is nearly full, at the point of junction. The common mode of connecting horse-shoe tile-drains is shown thus:

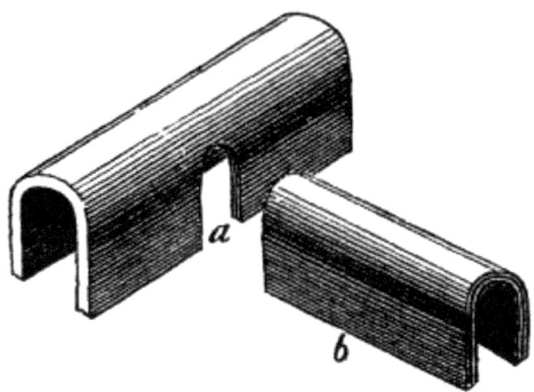

Fig. 50.—Junction of Drains.

Having no tiles made for the purpose, we, at first, formed the union by means of common hard bricks. Curving down the small drain toward the direction of the main, we left a space between two tiles of the main, of two or three inches, and brought down the last tile of the small drain to this opening, placing under the whole a flat stone, slate, or bricks, or a plank, to keep all firm at the bottom. Then we set bricks on edge on all sides, and covered the space [197] at the top with one or more, as necessary, and secured carefully against sand and the like.

We have since procured branch-pipes to be made at the tile-works, such as are in use in England, and find them much more satisfactory. The branches may be made to join the mains at any angle, and it might be advisable to make this part of both drains larger than the rest, to allow room for the obstructed waters to unite peacefully.

Fig. 51.
Branch Pipes.

The mains should be from three to six inches deeper than the minors. The fall from one to the other may usually be made most conveniently, by a gradual descent of three or four feet to the point of junction; but with branch-pipes, the fall may be nearly vertical, if desired, by turning the branch upward, to meet the small pipe. It will be necessary, in procuring branches for sole-tiles, to bear in mind that they are "rights and lefts," and must be selected accordingly, as the branch comes in upon the one or other side of the main.

The branch should enter the larger pipe not level with the bottom, but as high as possible, to give an inch fall to the water passing out of the branch into the main, to prevent possible obstruction at the junction.

## DRAINAGE INTO WELLS, OR SWALLOW HOLES.

In various parts of our country, there are lands lying too flat for convenient drainage in the ordinary methods, or too remote from any good outlet, or perhaps enclosed by lands of others who will not consent to an outfall through their domain, where the drainage water may be discharged into wells.

In the city of Washington, on Capitol Hill, it is a common practice to drain cellars into what are termed "dry wells." The surface formation is a close red clay, of a [198] few feet thickness, and then comes a stratum of coarse gravel; and the wells for water are sunk often as deep as sixty feet, indicating that the water-table lies very low. The heavy storms and showers fill the surface soil beyond saturation, and the water gushes out, literally, into the cellars and other low places. A dry well, sunk through the clay, conducts this water into the gravel bed, and this carries it away. This idea is often applied to land drainage. It is believed that there are immense tracts of fertile land at the West, upon limestone, where the surface might readily be relieved of surplus water, by conducting the mains into wells dug for the purpose. In some places, there are openings called "sink-holes," caused by the sinking of masses of earth, as in the neighborhood of the city of St. Louis, which would afford outlets for all the water that could be poured into them. In the Report of the Tioga County Agricultural Society for 1857, it is said in the *Country Gentleman*, that instances are given, where swamps were drained through the clay bottom into the underlying gravelly soil, by digging wells and filling them with stones.

In Fig. 7, at page 82, is shown a "fault" in the stratification of the earth; which faults, it is said, so completely carry off water, that wells cannot be sunk so as to reach it.

Mr. Denton says that in several parts of England, advantage is taken of the natural drainage existing beneath wet clay soils, by concentrating the drains to holes, called "swallow-holes." He says this practice is open to the objection that those holes do not always absorb the water with sufficient rapidity, and so render the drainage for a time, inoperative.

These wells are liable, too, to be obstructed in their operation by their bottoms being puddled with the clay carried into them by the water, and so becoming impervious. [199] This point would require occasional attention, and the removal of such deposits.

This principle of drainage was alluded to at the American Institute, February 14, 1859, by Professor Nash. He states, that there are large tracts of land having clay soil, with sand or gravel beneath the clay, which yet need drainage, and suggests that this may be effected by merely boring frequent holes, and filling them with pebbles, without ditches. In all such soils, if the mode suggested prove insufficient, large wells of proper depth, stoned up, or otherwise protected, might obviously serve as cheap and convenient outlets for a regular system of pipe or stone drains.

Mr. Bergen, at the same meeting, stated that such clayey soil, based on gravel, was the character of much of the land on Long Island; and we cannot doubt that on the prairies of the West, where the wells are frequently of great depth to obtain water for use, wells or swallow-holes to receive it, may often be found useful. Whenever the water-line is twenty or thirty feet below the surface, it is certain that it will require a large amount of water poured in at the surface of a well to keep it filled for any considerable length of time. The same principle that forces water into wells, that is, pressure from a higher source, will allow its passage out when admitted at the top.

We close this chapter with a letter from Mr. Denton. The extract referred to, has been here omitted, because we have already, in the chapter preceding this, given Mr. Denton's views, expressed more fully upon the same subject, with his own illustrations.

It should be stated that the letter was in reply to inquiries upon particular points, which, although disconnected, are all of interest, when touched upon by one whose opinions are so valuable. [200]

"London, 52 Parliament Street, Westminster, S. W.

"My Dear Sir:—I have received your letter of the 17th August, and hasten to reply to it.

"I am gratified at the terms in which you speak of my roughly-written 'Essays on Land Drainage.' If you have not seen my published letter to Lord Berners, and my recent essay 'On the Advantages of a Daily Record of Rain-fall,' I should much like you to look over them, for my object in both has been to check the uniformity of treatment which too much prevails with those who are officially called upon to direct draining, and who still treat mixed soils and irregular surfaces pretty much in the same way as homogeneous clays and even surfaces, the only difference being, that the distance between the drains is increased. We have now, without doubt, arrived at that point in the practice of draining in this country, which necessitates a revision of all the principles and rules which have been called into force by the Drainage Acts, and the institution of the Drainage Commission, whose duty it is to administer those Acts, and to protect the interests of Reversioners.

"This protection is, in a great measure, performed by the intervention of 'Inspectors of Drainage,' whose subordinate duty it is to see that the improvements provisionally sanctioned are carried out according to certain implied, if not fixed, rules. This is done by measuring depth and distance, which tends to a *parallel system (4 feet deep) in all soils*, which was Smith of Deanston's notion, only his drains were shallower, *i.e.*, from 2 to 3 feet deep.

"Some rules were undoubtedly necessary when the Commissioners first commenced dispensing the public money, and I do not express my objection to the absurd position to which these rules are bringing us, from any disrespect to them, nor with an idea that any better course could have been followed by the Government, in the first instance, than the adoption of the '*Parkes — Smith frequent drain system.*' This system was correctly applied, and continues to be correctly applied, to absorbent and retentive soils requiring the aeration of frequent drains to counteract their retentive nature; but it is altogether misapplied when adopted in the outcropping surfaces of the free water-bearing strata, which, though equally wet, are frequently drained by a comparatively few drains, at less than half the cost.

"The only circumstance that can excuse the indiscriminate adoption of a parallel system, is the fact, that all drains do some good, and the chances of a cure being greater in proportion to the number of drains, [201] it was not necessary to insist upon that judgment which ten years' experience should now give.

"My views on this point will perhaps be best understood by the following extract from an address I recently delivered. [Extract omitted, see p. 161].

* * * "I use one and a half inch pipes for the upper end of drains (*though I prefer two-inch*), one half being usually one and a half and the other half two-inch. This for minor drains; the mains run up to 9 or 10 inches, and even 18 inches in size, according to their service.

"There is no doubt sufficient capacity in one-inch pipes for minor drains; but, inasmuch as agricultural laborers are not mathematical scholars, and are apt to lay the pipes without precise junctions, it is best to have the pipes so large as to counteract that degree of carelessness which cannot be prevented. The ordinary price of pipes in this country will run thus: + meaning *above*, and-*below*, the prices named:

| 1½ | inch | 15s. | + |
|----|------|------|---|
| 2 | " | 20s. | - |
| 3 | " | 30s. | |
| 4 | " | 40s. | + |
| 5 | " | 50s. | + |
| 6 | " | 60s. | + |

"The price of cutting clays 4 feet deep, will vary from 1d. to 1½d. per yard, according to density and mixture with stone; and the price of cutting in mixed soils will vary from 1½d. to 6d., according to the quantity of pick-work and rock, and with respect, also, to the price of agricultural labor. (See my tabular table of cost in Land Drainage and Drainage Systems.)

"I should have thought it would have been quite worth the while of the American Government to have had a farm of about 500 acres, drained by English hands, under an experienced engineer, as a practical sample of English work, for the study of American agriculturists, with every drain laid down on a plan, with the sizes of the pipes, and all details of soil, and prices of labor and material, set forth.

"I am, dear Sir,
"Yours very faithfully,
"The Hon. H. F. French, Exeter. "J. BAILEY DENTON."

# CHAPTER IX [202]
# THE COST OF TILES—TILE MACHINES.

Prices far too high; Albany Prices.—Length of Tiles.—Cost in Suffolk Co., England.—Waller's Machine.—Williams' Machine.—Cost of Tiles compared with Bricks.—Mr. Denton's Estimate of Cost.—Other Estimates.—Two-inch Tiles can be Made as Cheaply as Bricks.—Process of Rolling Tiles.—Tile Machines.—Descriptions of Daines'.—Pratt & Bro.'s.

The prices at which tiles are sold is only, as the lawyers say, *primâ facie* evidence of their cost. It seems to us, that the prices at which tiles have thus far been sold in this country, are very far above those at which they may be profitably manufactured, when the business is well understood, and pursued upon a scale large enough to justify the use of the best machinery. The following is a copy of the published prices of tiles at the Albany Tile Works, and the same prices prevail throughout New England, so far as known:

| Horse-shoe Tile--Pieces. | | | | | Sole-Tile--Pieces. | | | | |
|---|---|---|---|---|---|---|---|---|---|
| 2½ | inches | rise | $12 | per 1000. | 2 | inches | rise | $12 | per 1000. |
| 3½ | " | " | 15 | " | 3 | " | " | 18 | " |
| 4½ | " | " | 18 | " | 4 | " | " | 40 | " |
| 5½ | " | " | 40 | " | 5 | " | " | 60 | " |
| 6½ | " | " | 60 | " | 6 | " | " | 80 | " |
| 7½ | " | " | 75 | " | 8 | " | " | 125 | " |

Few round pipe-tiles have yet been used in this country, although they are the kind generally preferred by engineers in England. The prices of round tiles would vary little from those of sole-tiles.

Tiles are usually cut fourteen inches long, and shorten, [203] in drying and burning, to about twelve and a half inches, so that, with breaking and other casualties, they may be calculated to lay about one foot each; that is to say, 1,000 tiles may be expected to lay 1,000 feet of drains.

To assist those who desire to manufacture tiles for sale, or for private use, it is proposed to give such information as has been gathered from various sources as to the cost of making, and the selling prices of tiles, in England. The following is a memorandum made at the residence of Mr. Thomas Crisp, at Butley Abbey, in Suffolk Co., Eng., from information given the author on the 8th of July, 1857:

"Mr. Crisp makes his own tiles, and also supplies his neighbors who need them. He sells one and a half inch pipes at 12s. ($3) per 1,000. He pays 5s. ($1.25) per 1,000 for having them made and burnt. His machine is Waller's patent, No. 22, made by Garrett and Son, Leiston, Saxemundham, Suffolk. It works by a lever, makes five one and a half inch pipes at once, or three sole-tiles about two-inch. The man at work said, that he, with a man to carry away, &c., could make 4,000 one and a half inch pipes per day. They used no screen, but cut the clay with a wire. The machine cost £25 (about $125). At the kiln, which is permanent, the tiles are set on end, and bricks with them in the same kiln. They require less heat than bricks, and *cost about half as much* as bricks here, which are moulded ten inches by five.

"Two girls were loading bricks into a horse-cart, and two women receiving them, and setting them in the kiln. They made roof-tiles with the same machine, and also moulded large ones by hand. The wages of the women are about 8d. (sixteen cents) per day."

At the exhibition of the Royal Agricultural Society, in England, the author saw Williams' Tile Machine in operation, and was there informed by the exhibitor, who said [204] he was a tile-maker, that it requires *five-sevenths as much coal* to burn 1,000 two-inch tiles, as 1,000 bricks—the size of bricks being 10 by 5; and he declared, that he, with one boy, could make with the machine, 7,000 two-inch tiles per day, after the clay is prepared. Of course, one other person, at least, must be employed to carry off the tiles.

Mr. Denton gives his estimates of the prices at which pipe-tiles may be procured in England, as follows—the prices, which he gives in English currency, being translated into our own:

"When ordinary agricultural labor is worth $2 50 per week, pipes half one and a half inch, and half two-inch, maybe taken at an average cost of $4 38 per 1,000. When labor is $3 00 per week, the pipes will average $5 00 per 1,000, and when labor is $3 50, they will rise to $5 62."

He adds: "In giving the above average cost of materials, those districts are excluded from consideration, where clay suitable for pipes, exists in the immediate vicinity of coal-pits, which must necessarily reduce the cost of producing them very considerably."

Taking the averages of several careful estimates of the cost of tiles and bricks, from the "Cyclopædia of Agriculture," we have the price of tiles in England about $5 per 1,000, and the price of bricks $7.87, from which the duty of 5s. 6d. should be deducted, leaving the average price of bricks $6.50. Upon tiles there is no such duty. Bricks in the United States are made of different sizes, varying from 8 × 4 in. to the English standard 10 × 5 in. Perhaps a fair average price for bricks of the latter size, would be not far from $5 per 1,000; certainly below $6.50 per 1,000. There is no reason why tiles may not be manufactured in the United States, as cheaply, compared with the prices of bricks, as in England; and it is quite clear that tiles of the sizes named, are far cheaper there than common bricks.

What is wanted in this country is, first, a demand sufficient [205] to authorize the establishment of works extensive enough to make tiles at the best advantage; next, competent skill to direct and perform the labor; and, finally, the best machinery and fixtures for the purpose. It is confidently predicted, that, whenever the business of tile-making becomes properly established, the ingenuity of American machinists will render it easy to manufacture tiles at English prices, notwithstanding the lower price of labor there; and that we shall be supplied with small tiles in all parts of the country at about the current prices of bricks, or at about one half the present Albany prices of tiles, as given at the head of this chapter. It should be mentioned here, perhaps, that, in England, it is common to burn tiles and bricks together in the same kiln, placing the tiles away from the hottest parts of the furnace; as, being but about half an inch in thickness, they require less heat to burn them than bricks.

In the estimates of labor in making tiles in England, a small item is usually included for "rolling." Round pipes are chiefly used in England. When partly dried, they are taken up on a round stick, and rolled upon a small table, to preserve their exact form. Tiles usually flatten somewhat in drying, which is not of importance in any but round pipes, but those ought to be uniform. By this process of rolling, great exactness of shape, and a great degree of smoothness inside, are preserved.

## TILE MACHINES.

Drainage with tiles is a new branch of husbandry in America. The cost of tiles is now a great obstacle in prosecuting much work of this kind which land-owners desire to accomplish. The cost of tiles, and so the cost of drainage, depends very much—it may be said, chiefly—upon the perfection of the machinery for tile-making; and here, as almost everywhere else, agriculture and the [206] mechanic arts go hand in hand. Labor is much dearer in America than in Europe, and there is, therefore, more occasion here than there, for applying mechanical power to agriculture. We can have no cheap drainage until we have cheap tiles; and we can have cheap tiles only by having them made with the most perfect machinery, and at the lowest prices at which competing manufacturers, who understand their business, can afford them.

In the preceding remarks on the *cost of tiles*, may be found estimates, which will satisfy any thinking man that tiles have not yet been sold in America at reasonably low prices.

To give those who may desire to establish tileries, either for public or private supply, information, which cannot readily be obtained without great expense of English books, as to the prices of tile machines, it is now proposed to give some account of the best English machines, and of such American inventions as have been brought to notice.

It is of importance that American machinists and inventors should be apprised of the progress that has been made abroad in perfecting tile machines; because, as the subject attracts attention, the ingenuity of the universal Yankee nation will soon be directed toward the discovery of improvements in all the processes of tile-making. Tiles were made by hand long before tile machines were invented.

A Mr. Read, in the "Royal Agricultural Journal," claims to have used *pipe* tiles as early as 1795, made by hand, and formed on a round stick. No machine for making tiles is described, before that of Mr. Beart's, in 1840, by which "common tile and sole (not pipes or tubes) were made." This machine, however, was of simple structure, and not adapted to the varieties of tiles now used. [207]

All tile machines seem to operate on the same general principle — that of forcing wet clay, of the consistency of that used in brick-making, through apertures of the desired shape and size. To make the mass thus forced through the aperture, *hollow,* the hole must have a piece of metal in the centre of it, around which the clay forms, as it is pushed along. This centre piece is kept in position by one or two thin pieces of iron, which of course divide the clay which passes over them, but it unites again as it is forced through the die, and comes out sound, and is then cut off, usually by hand, by means of a small wire, of the required length, about fourteen inches.

Tile machines work either vertically or horizontally. The most primitive machine which came to the author's notice abroad, was one which we saw on our way from London to Mr. Mechi's place. It was a mere upright cylinder, of some two feet height, and perhaps eight inches diameter, in which worked a piston. The clay was thrown into the cylinder, and the piston brought down by means of a brake, like an old-fashioned pump, and a single round pipe-tile forced out at the bottom. The force employed was one man and two boys. One boy screened the clay, by passing through it a wire in various directions, holding the wire by the ends, and cutting through the mass till he had found all the small stones contained in it. The man threw the masses thus prepared, into the cylinder, and put on the brake, and the other boy received the tiles upon a round stick, as they came down through the die at the bottom, and laid them away. The cylinder held clay enough to make several, perhaps twenty, two-inch pipes. The work was going on in a shed without a floor, and upon a liberal estimate, the whole establishment, including shed and machine, could not cost more than fifty dollars. Yet, on this simple plan, tiles were moulded much more rapidly than bricks were [208] made in the same yard, where they were moulded singly, as they usually are in England. It was said that this force could thus mould about 1,800 small tiles per day.

This little machine seems to be the same described by Mr. Parkes as in general use in 1843, in Kent and Suffolk Counties.

Most of the tile machines now in use in England and America, are so constructed, as to force out the tiles upon a horizontal framework, about five two-inch, or three three-inch pipes abreast. The box to contain the clay may be upright or horizontal, and the power may be applied to a wheel, by a crank turned by a man, or by horse, steam, or water power, according to the extent of the works.

We saw at the Exhibition of the Royal Agricultural Society, at Salisbury, in England, in July, 1857, the "pipe and tile machine," of W. Williams, of Bedford. It was in operation, for exhibition, and was worked by one man, who said he was a tile maker, and that he and one boy could make with the machine 7,000 two-inch tiles per day, after the clay was prepared in the pug mill. Four tiles were formed at once, by clay passed through four dies, and the box holds clay enough for thirty-two two-inch tiles, so that thirty-two are formed as quickly as they can be removed, and as many more, as soon as the box can be refilled.

The size, No. 3, of this machine, such as we then saw in operation, and which is suitable for common use, costs at Bedford $88.50, with one set of dies; and the extra dies, for making three, four, and six-inch pipes, and other forms, if desired, with the *horses*, as they are called, for removing the tiles, cost about five dollars each.

This, like most other tile machines, is adapted to making tiles for roofs, much used in England instead of shingles or slates, as well as for draining purposes.

There are several machines now in use in England [209] namely: Etheridge's, Clayton's, Scragg's, Whitehead's, and Garrett's—either of which would be satisfactory, according to the amount of work desired.

We have in America several patented machines for making tiles, of the comparative merits of which we are unable to give a satisfactory judgment. We will, however, allude to two or three, advising those who are desirous to purchase, to make personal examination for themselves. We are obliged to rely chiefly on the statements of the manufacturers for our opinions.

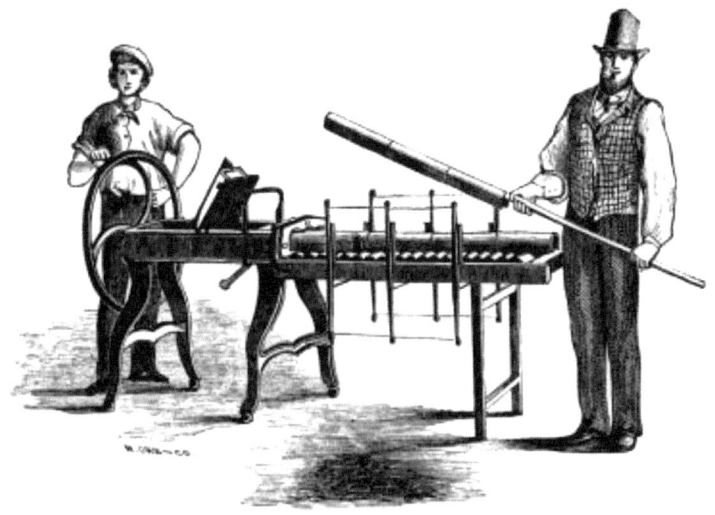

DAINES' DRAIN TILE MAKER

Daines' American Drain Tile Machine is manufactured at Birmingham, Michigan, by John Daines. This machine is in use in Exeter, N. H., close by the author's residence, and thus far proves satisfactory. The price of it is about $100, and the weight, about five hundred pounds. It occupies no more space than a common three-and-a-half foot table, and is worked by a man at a crank. It is capable of turning out, by man power, about two hundred and fifty two-inch tiles in an hour, after the clay is prepared in a pug mill. Horse or water power can be readily attached to it.

We give a drawing of it, not because we are sure it is the best, but because we are sure it is a good machine, and to illustrate the principle upon which all these machines are constructed.

Pratt's Tile Machine is manufactured at Canandaigua, New York, by Pratt & Brothers, and is in use in various places in that State as well as elsewhere. This machine differs from Daines' in this essential matter, that here the clay is *pugged*, or tempered, and formed into tiles at one operation, while with Daines' machine, the clay is first passed through a pug mill, as it is for making bricks in the common process.

Pratt's machine is worked by one or two horses, or by steam or water power, as is convenient. The price of the smaller size, worked by one horse, is $150, and the price [210] of the larger size, worked by two horses, $200. Professor Mapes says he saw this machine in operation and considers it "perfect in all its parts." The patentees claim that they can make, with the one-horse machine, 5,000 large tiles a day. They state also that "two horses will make tiles about as cheap as bricks are usually made, and as fast, with the large-sized machine."

Fig. 53.—Pratt's Tile Machine.

These somewhat indefinite statements are all that we can give, at present, of the capacity of the machines. We should have no hesitation in ordering a Pratt machine were we desirous of entering into an extensive business of Tile-making, and we should feel quite safe with a Daines' machine for a more limited manufacture.

## SALISBURY'S TILE MACHINE.

S. C. Salisbury, at the Novelty Works, in the city of New York, is manufacturing a machine for making tiles and bricks, which exhibits some new and peculiar features, worthy of attention by those who propose to purchase tile machines. Prof. Mapes expresses the confident opinion that this machine excels all others, in its capacity to form tiles with rapidity and economy. We have examined only a working model. It is claimed that the large size, with horse-power, will make 20,000 two-inch tiles per day, and the hand-power machine 3,000 per day. We advise tile makers to examine all these machines in operation, before purchasing either.

# CHAPTER X [211]
# THE COST OF DRAINAGE.

Draining no more expensive than Fencing.—Engineering.—Guessing not accurate enough.—Slight Fall sufficient.—Instances.—Two Inches to One Thousand Feet.—Cost of Excavation and Filling.—Narrow Tools required.—Tables of Cubic contents of Drains.—Cost of Drains on our own Farm.—Cost of Tiles.—Weight and Freight of Tiles.—Cost of Outlets.—Cost of Collars.—Smaller Tiles used with Collars.—Number of Tiles to the Acre, with Tables.—Length of Tiles varies.—Number of Rods to the Acre at different Distances.—Final Estimate of Cost.—Comparative Cost of Tile-Drains and Stone-Drains.

A prudent man, intending to execute a work, whether it be "to build a tower," or drain a field, "sitteth down first and counteth the cost, whether he hath sufficient to finish it." There is good sense and discretion in the inquisitiveness which suggests so often the inquiry, "How much does it cost to drain an acre?" or, "How much does it cost a rod to lay drains?" These questions cannot be answered so briefly as they are asked; yet much information can be given, which will aid one who will investigate the subject.

The process of drainage is expensive, as compared with the price of land in our new settlements; but its cost will not alarm those who have been accustomed to see the improvements made in New England upon well cultivated farms. Compared with the labor and cost of building and maintaining fences upon the highways, and in the subdivisions of lots, common in the Eastern States, the [212] drainage of land is a small matter. We see in many places long stretches of faced walls, on the line of our roads near towns and villages, which cost from two to five dollars per rod. Our common "stone walls" in these States cost about one dollar per rod to build originally; and almost any kind of wooden fence costs as much. Upon fences, there is occasion for annual repairs, while drains properly laid, are permanent.

These suggestions are thrown out, that farmers may not be alarmed without cause, at the high cash estimates of the cost of drainage operations. Money comes slowly to farmers, and a cash estimate looks larger to them than an estimate in labor. The cost of fencing seems no great burden; though, estimated in cash, it would seem, as in fact it is, a severe charge.

Drainage can be performed principally by the same kind of labor as fencing, the cost of the tiles being a small item in the whole expense. The estimates of labor will be made at one dollar per day, in investigating this matter.

This would be the fair cash value of work by the day, perhaps; but it is far more than farmers, who have work in hand on their own farms, which may be executed in the leisure season after haying, and even into the Winter, when convenient, will really expend for such labor. Few farm operations would pay expenses, if every hour of superintendence, and every hour of labor by man and boy and beast, were set down at this high rate.

The cost of the tiles will, ordinarily, be a cash item, and the labor may be performed like that of planting, hoeing, haying, and harvesting, by such "help" hired by the mouth or day, or rendered by the family, as may be found convenient.

The cost of drainage may be considered conveniently, to borrow a clerical phrase, "under the following heads." [213]

1. *Laying out, or Engineering.* — In arranging our Spring's work, we devote time and attention to laying it out, though this hardly forms an item in the expense of the crop. Most farmers may think themselves competent to lay out their drainage-works, without paying for the scientific skill of an engineer, or even of a surveyor.

It is believed, however, that generally, it will be found true economy, to procure the aid of an experienced engineer, if convenient, to lay out the work at the outset. Certainly, in most cases, some skill in the use of levelling instruments, at least, is absolutely essential to systematic work. No man, however experienced, can, by the eye, form any safe opinion of the fall of a given tract of land. Fields which appear perfectly level to the eye, will be found frequently to give fall enough for the deepest drainage. The writer recently had occasion to note this fact on his own land.

A low wet spot had many times been looked at, as a place which should be drained, both to improve its soil, and the appearance of the land about it; but to the eye, it seemed doubtful whether it was not about as low as the stream some forty rods off, into which it must be drained. Upon testing the matter carefully with levelling instruments, it was found that from the lowest spot in this little swamp, there was a fall of seven and a half feet to the river, at its ordinary height! Again, there are cases where it will be found upon accurate surveys, that the fall is very slight, so that great care will be requisite, to lay the drains in such a way that the descent may be continuous and uniform.

Without competent skill in laying out the work, land-owners will be liable not only to errors in the fall of the drains, but to very expensive mistakes in the location of them. A very few rods of drains, more than are necessary, [214] would cost more than any charge of a competent person for laying them out properly.

Again, experience gives great facility in judging of the underground flow of water, of the permeability of soil, of the probability of finding ledges or other rock formation, and many other particulars which might not suggest themselves to a novice in the business.

The laying out of drains is important, not only with reference to the work in hand, but to additional work to be executed in future on adjoining land, so that the whole may be eventually brought into one cheap and efficient system with the smallest effective number of drains, both minors and mains, and the fewest outlets possible; with such wells, or other facilities for inspection, as may be necessary.

In the English tables of the cost of drainage by the Drainage Companies, an estimate of $1.25 per acre is usually put down for "superintendence," which includes the engineering and the supervision of the whole process of opening, laying and filling, securing outfalls, and every other process till the work is completed. The general estimate of the cost of drainage is about $25.00 per acre, and this item of $1.25 is but a small per centage on that amount. The point has been dwelt upon here, more for the purpose of impressing upon land-owners, the importance of employing competent skill in the laying out of their drainage works, than because the expense thus incurred, forms any considerable item of the cost of the whole work.

2. *Excavation and Filling.* The principal expense of drainage is incurred in the excavation of the ditch, whether it be for tiles or for stones. The labor of excavation depends much upon the nature of the soil to be moved.

"Draining on a sound clay," says the writer of a prize essay, "free [215] from stones, may be executed at a cheaper rate per rod, in length, than on almost any other kind of soil, as, from the firmness of the clay, the work may be done with narrow spades, and but a small quantity of soil requires to be removed. The draining of wet sands or grounds, or clays in which veins of sand abound, is more expensive than on sound clays, because a broader spade has to be used, and consequently a larger amount of soil removed; and draining stony or rocky soils is still more expensive, because the pick has to be used. This adds considerably to the expense."

Great stress is laid, by all experienced persons, upon using narrow spades, and opening ditches as narrow as possible.

It is somewhat more convenient for unskillful laborers to work in a wide ditch than in a narrow one, and although the laborers frequently protest that they cannot work so rapidly in narrow ditches, yet it is found that, in contract work, by the rod, they usually open the ditches very narrow.

Indeed, it will be found that, generally, the cost of excavation bears a pretty constant proportion to the number of cubic feet of earth thrown out.

It will surprise those unaccustomed to these estimates, to observe how rapidly the quantity excavated, increases with the increased width of the ditch.

To enable the reader accurately to compute the measurement of drains of any dimensions likely to be adopted, a table and explanations, found in the Report of the Board of Health, already quoted, are given below. The dimensions, or contents of any drain, are found by multiplying together the length, depth, and *mean* width of the drain.

"Thus, if a drain is 300 yards long, and the cutting 3 feet deep, 20 inches wide at the top, and 4 inches wide at the bottom, the mean width would be 12 inches (or the half of the sum of 20 and 4), and if we multiply 300, the length, by 1, the depth in yards, and by 1/3, the mean width in yards, and the product would be 100 cubic yards. The following table will serve to facilitate such calculations.

[216] Table showing the number of Cubic Yards of Earth in each Rod (5½ Yards in length), in Drains or Ditches of various Dimensions.

**Depth.**                                        **Mean Width.**

| Inches. | 7 In. | 8 In. | 9 In. | 10 In. | 11 In. | 12 In. | 13 In. | 14 In. | 15 In. | 16 In. | 17 In. | 18 In. |
|---|---|---|---|---|---|---|---|---|---|---|---|---|
| 30 | 0.89 | 1.02 | 1.146 | 1.27 | 1.40 | 1.53 | 1.655 | 1.78 | 1.91 | 2.04 | 2.164 | 2.29 |
| 33 | 0.98 | 1.12 | 1.26 | 1.40 | 1.54 | 1.68 | 1.82 | 1.96 | 2.10 | 2.24 | 2.38 | 2.52 |
| 36 | 1.07 | 1.22 | 1.375 | 1.53 | 1.68 | 1.83 | 1.986 | 2.14 | 2.29 | 2.244 | 2.60 | 2.75 |
| 39 | 1.16 | 1.324 | 1.49 | 1.655 | 1.82 | 1.986 | 2.15 | 2.32 | 2.48 | 2.65 | 2.81 | 2.98 |
| 42 | 1.25 | 1.426 | 1.604 | 1.78 | 1.96 | 2.14 | 2.32 | 2.495 | 2.674 | 2.85 | 3.03 | 3.21 |
| 45 | 1.34 | 1.53 | 1.72 | 1.91 | 2.10 | 2.29 | 2.48 | 2.67 | 2.865 | 3.055 | 3.246 | 3.438 |
| 48 | 1.426 | 1.63 | 1.833 | 2.04 | 2.24 | 2.444 | 2.65 | 2.85 | 3.056 | 3.26 | 3.46 | 3.667 |
| 51 | 1.515 | 1.73 | 1.95 | 2.164 | 2.38 | 2.60 | 2.81 | 3.03 | 3.25 | 3.46 | 3.68 | 3.896 |
| 54 | 1.604 | 1.83 | 2.06 | 2.29 | 2.52 | 2.75 | 2.98 | 3.20 | 3.44 | 3.666 | 3.895 | 4.125 |
| 57 | 1.69 | 1.935 | 2.18 | 2.42 | 2.66 | 2.90 | 3.14 | 3.38 | 3.63 | 3.87 | 4.11 | 4.354 |
| 60 | 1.78 | 2.036 | 2.29 | 2.546 | 2.80 | 3.056 | 3.31 | 3.564 | 3.82 | 4.074 | 4.33 | 4.584 |

"Along the top of the table is placed the mean widths in inches, and on the left-hand side the depths of the drains, extending from 30 inches to 5 feet. The numbers in the body of the table express cubic yards, and decimals of a yard. In making use of the table, it is necessary first to find the mean width of the drain, from the widths at the top and bottom. Thus, if a drain 3 feet deep were 16 inches wide at the top, and 4 inches at the bottom, the mean width would be half of 16 added to 4, or 10; then, by looking in the table for the column under 10 (width), and opposite 36 (inches of depth), we find the number of cubic yards in each rod of such a drain to be 1.53, or somewhat more than one and a half. If we compare this with another drain 20 inches wide at the top, 4 inches at the bottom, and 4½ feet deep, we have the mean width 12, and looking at the table under 12 and opposite 54, we find 2.75 cubic yards, or two and three-quarters to the rod. In this case, the quantity of earth to be removed is nearly twice as much as in the other, and hence, as far as regards the digging, the cost of the labor will be nearly double. But in the case of deep drains, the cost increases slightly for another reason, namely, the increased labor of lifting the earth to the surface from a greater depth."

Under the title of the "Depth of Drains," other reasons are suggested why shallow drains are more easily wrought than deeper drains. The widths given in English treatises, and found perfectly practicable there, with proper drainage-tools, will seem to us exceedingly narrow. Mr. Parkes gives the width of the top of a four-foot drain 18 [217] inches, of a three-and-a-half foot drain 16 inches, and of a three-foot drain 12 inches. He gives the width of drains for tiles, three inches at bottom, and those for stones, eight inches. Of the cost of excavating a given number of cubic yards of earth from drains, it is difficult to give reliable estimates. In the writer's own field, where a pick was used to loosen the lower two feet of earth, the labor of opening and filling drains 4 feet deep, and of the mean width of 14 inches, all by hand labor, has been, in a mile of drains, being our first experiments, about one day's labor to three rods in length. The excavated earth of such a drain, measures not quite three cubic yards. (Exactly, 2.85.)

In work subsequently executed, we have opened our drains of 4 foot depth, but 20 inches at top, and 4 inches at bottom, giving a mean width of 12 inches. In one instance, in the Summer of 1858, two men opened 14 rods of such drain in one day. In six days, the same two men opened, laid, and filled 947 feet, or about 57½ rods of such drain. Their labor was worth $12.00, or 21 cents per rod. The actual cost of this job was as follows:

| | |
|---|---:|
| 847 two-inch tiles, at $13 per 1,000 | $11.01 |
| 100 three-inch tiles, at $13 per 1,000 for main | 2.50 |
| 70 bushels of tan, to protect the joints | .70 |
| Horse to haul tiles and tan | .50 |
| Labor, 12 days, at $1 | 12.00 |
| Total | $26.71 |

This is 46½ cents per rod, besides our own time and skill in laying out and superintending the work. The work was principally done with Irish spades, and was in a sandy soil. In the same season, the same men opened, laid, and filled 70 rods of four-foot drain, of the same mean width of 12 inches, in the worst kind of clay soil, where the pick was constantly used. It cost 35 days' labor to complete the job, being 50 cents per rod for the labor alone. The least cost of the labor of draining 4 feet deep, on our [218] own land, is thus shown to be 21 cents per rod, and the greatest cost 50 cents per rod, all the labor being by hand. One-half these amounts would have completed the drains at 3 feet depth, as has been already shown.

But the excavation here is much greater than is usual in England, Mr. Parkes giving the mean width of a four-foot drain but 10½ inches, instead of 14 or 12, as just given. Mr. Denton gives estimates of the cost, in England, of cutting and filling four-foot drains, which vary from 12 cents per rod upwards, according to the prices of labor, and other circumstances.

In New England, where labor may be fairly rated at one dollar per day, the cost of excavating and filling four-foot drains by hand labor, must vary from 20 to 50 cents per rod, according to the soil, and half those amounts for drains of three-foot depth.

Of the aid which may be derived from the use of draining plows, or of the common plow, or subsoil plow, our views may be found expressed under the appropriate heads. That drains will long continue to be opened in this vast country by hand labor, is not to be |

supposed, but we give our estimates of the expenses, at this first stage of our education in drainage.

3. *Cost of the Tiles.* Under the title of "The Cost of Tiles," we have given such information as can be at present procured, touching that matter. It will be assumed, in these estimates, that no tiles of less than 1½ inch bore will be used for any purpose, and for mains, usually those of three-inch bore are sufficient. The proportion of length of mains to that of minors is small, and, considering the probable reduction of prices, we will, for the present, assume $10 per 1,000 as the prices of such mixed sizes as may be used.

Add to this, the freight of them to a reasonable distance, and we have the cost of the tiles on the field. The [219] weight of two-inch tiles is usually rated at about 3 lbs. each, though they fall short of this weight until wet.

4. *Outlets.* A small per-centage should be added to the items already noticed, for the cost of the general outfall, which should be secured with great care; although, from such examination as the writer has made in this country, and in England also, in the large majority of cases, drains are discharged with very little precaution to protect the outlets. Works completed under the charge of regular engineers, form an exception to this remark; and an item of 37 cents per acre, for iron outlets and masonry, is usually included in the estimated cost per acre of drainage.

5. *Collars.* It is not known to the author that collars have been at all used in America, except at the New York Central Park, in 1858; round pipes, upon which they are commonly used abroad, when used on any, not being yet much in use here.

In the estimates of Mr. Denton, in his tables, collars are set down at about half the cost of the mixed tiles. The bore of them being large enough to receive the end of the tile, increases the price in proportion to the increase in size. It is believed, however, that a smaller size of tiles may prudently be used with collars than without, because the collars keep the tiles perfectly in line, and freely admit water, while they exclude roots, sand, and other obstructions. A drain laid with one and a half inch tiles with collars is, no doubt, better in any soil than two-inch tiles without collars. Some compensation for the cost of collars may thus be found in the less price of the smaller tiles.

6. *Laying.* The cost of laying tiles is so trifling as hardly to be worth estimating, except to show its insignificance. The estimate, by English engineers, is two cents per rod for "pipe laying and finishing." What is included in "finishing," does not appear. From the personal observations of the writer, it is believed that an [220] active man may lay from 60 to 100 rods of tiles per day, in ditches well prepared. Indeed, we have seen our man James, lay twelve rods of two-inch tiles, in a four-foot ditch, in forty-five minutes, when he was not aware that he was working against time. This is at the rate of sixteen rods an hour, which would give just 160 rods, or a half-mile, in a day of ten hours.

7. *Number of Tiles to the Acre.* The number of tiles used depends, of course, upon the distances apart of the drains, and upon the length of the tiles used.

The following table gives the number of tiles of various length, per acre, required at different intervals:

| Intervals be-tween the Drains, in feet. | Twelve inch Pipe. | Thirteen inch Pipe. | Fourteen inch Pipe. | Fifteen inch Pipe. |
|---|---|---|---|---|
| 15 | 2904 | 2680 | 2489 | 2323 |
| 18 | 2420 | 2234 | 2074 | 1936 |
| 21 | 2074 | 1915 | 1778 | 1659 |
| 24 | 1815 | 1676 | 1555 | 1452 |
| 27 | 1613 | 1489 | 1383 | 1290 |
| 30 | 1452 | 1340 | 1244 | 1161 |
| 33 | 1320 | 1219 | 1131 | 1056 |
| 36 | 1210 | 1117 | 1037 | 968 |

| | | | |
|---|---|---|---|
| 39 | 1117 | 1031 | 957 | 893 |
| 42 | 1037 | 958 | 888 | 829 |

The following table gives the number of rods per acre of drains at different distances:

| Intervals between the Drains, in feet. | Rods per acre. |
|---|---|
| 15 | 176 |
| 18 | 146-2/3 |
| 21 | 125-5/7 |
| 24 | 110 |
| 27 | 97-7/9 |
| 30 | 88 |
| 33 | 80 |
| 36 | 73-1/3 |
| 39 | 67-9/13 |
| 42 | 62-6/7 |

[221] It may be remarked here, that tiles, moulded of the same length, vary nearly two inches when burned, according to the severity of the heat. It may be suggested, too, that the length of the tile, in the use of any machine, is entirely at the option of the maker. It is not, perhaps, an insult to our common humanity, to suggest to buyers the propriety of measuring the length as well as calibre of tiles before purchasing. In the estimates which will be made in this detail, it will be assumed that tiles will lay one foot each, with allowance for imperfections and breakage. This is as near as possible to accuracy, according to our best observation; and, besides, there is convenience in this simple estimate of one tile to one foot, which is important in practice.

We have now the data from which we may make some tolerably safe estimates of the cost of drainage. With labor at one dollar per day, and tiles at $10 per 1,000, or one cent each, or one cent a foot, and ditches four feet deep, opened and filled at one-third of a day's labor to the rod, we may set down the principal items of the cost of drainage by the rod, as follows:

| | | |
|---|---|---|
| Cutting and filling per rod | 33⅓ | cts. |
| Tiles | 16⅔ | " |
| | 50 | |

This is putting the tiles at one cent a foot, and the labor at two cents a foot, or just twice as much as the cost of tiles, and it brings a total of half a dollar a rod, all of them numbers easily remembered, and convenient for calculation.

By reference to the table giving the number of rods to the acre, the cost of labor and tiles per acre may be at once found, by taking half the number of rods in dollars. At 42 feet distance, the cost will be $31.42 per acre; at [222] 30 feet distance, $44; and at 60 feet, half that amount, or $22 per acre.

Our views as to the frequency of drains, may be found under the appropriate head.

Our estimate thus far, is of four-foot drains. We have shown, under the head of the "Depth of Drains," that the cost of cutting and filling a four-foot drain is double that of cutting and filling a three-foot drain. There is no doubt, that, after all the good advice we have given on this subject, many, who "grow wiser than their teachers are," will set aside the teachings of the best draining engineers in the world, and insist that three feet deep is enough, and persist in so laying their tiles.

This *shallowness* will reduce the cost of labor about one half, so that we shall have the cost of labor and tiles equal—one cent a foot, making 33? cents per rod, or one-third of a dollar, instead of one-half a dollar per rod. To the cost of labor and tiles, we should add a fair estimate of the cost of the other items of engineering and outlets. These are trifling matters, which English tables, as has been shown, estimate together, at about $1.67 per acre.

Briefly to recapitulate the elements of computation of the cost of drainage, we find them to be these: the price of labor, the price of tiles, and freight of them; the character of the soil, the depth of the drains, and their distance apart, with the incidental expense of engineering and of outfalls, and the large additional cost of *collars*, where they are deemed necessary.

## COMPARATIVE COST OF TILE AND STONE DRAINS.

It is not possible to answer, with precision, the question so often asked, as to the comparative cost of drainage with tiles and stones.

The estimates given of the cost of tile drains, are based upon the writer's own experience, upon his own farm [223] mainly; and the mean width of four-foot tile drains, may be assumed to be 14 inches, instead of 10½ inches, as actually practiced in England.

For a stone drain of almost any form, certainly for any regular water-course laid with stones, our ditch must be at least 21 inches wide from top to bottom. This is just 50 per cent, more than our own estimate, and 100 per cent., or double the English estimate for tile drains.

It will require at least two ox-cart loads of stones to the rod, to construct any sort of a stone drain, costing, perhaps, 25 cents a load for picking up and hauling. In most cases, where the stones are not on the farm, it will cost twice that sum. We will say 25 cents per rod for laying the stones, though this is a low estimate. We have, then, for cutting and filling the ditch, 50 cents per rod, 50 cents for hauling stone, and for laying, 25 cents per rod, making $1.25 a rod for a stone drain, against 50 cents per rod for tile drains.

Then we have a large surplus of earth, two cartloads to the rod, displaced by the two loads of stone, to be disposed of; and in case of the tiles, we have just earth enough. There are many other considerations in favor of tiles: such as the cutting up of the ground by teaming heavy loads of stones; the greater permanency of tiles; and the fact that they furnish no harbor for mice and other vermin, as the

English call such small beasts. In favor of stones, is the fact, that often they are on the land, and must be moved, and it is convenient to dispose of them in the ditches.

Again, there are many parts of the country where tiles are not to be procured, without great cost of freight, and where labor is abundant at certain seasons, and money scarce at all seasons, so that the question is really between stone drains and no drains.

Stone drains, if laid very deep, are far more secure than [224] when shallow; because, if shallow, they are usually ruined by the breaking in of water at the top, in the Spring time, by the action of frost, and by the mining of mice and moles. If laid four feet deep, and the earth rammed hard above the stones, and rounded on the surface to throw off surface water, they may be found efficient and permanent.

The conclusion, however, is, that where it can be procured, at any reasonable cost, drainage with tiles will generally cost less than one-half the expense of drainage with stones, and be incomparably more satisfactory in the end.

# CHAPTER XI [225]
# DRAINING IMPLEMENTS.

Unreasonable Expectations about Draining Tools. — Levelling Instruments; Guessing not Accurate. — Level by a Square. — Spirit Level. — Span, or A Level. — Grading by Lines. — Boning-rod. — Challoner's Drain Level. — Spades and Shovels. — Long-handled Shovel. — Irish Spade, Description and Cut. — Bottoming Tools. — Narrow Spades. — English Bottoming Tools. — Pipe-layer. — Pipe-laying Illustrated. — Pick-axes. — Drain Gauge. — Drain Plows, and Ditch-Diggers. — Fowler's Drain Plow. — Pratt's Ditch-Digger. — McEwan's Drain Plow. — Routt's Drain Plow.

It seems to be a characteristic of Americans, to be dissatisfied with every recent improvement in art or science, and the greater the step in advance of former times, the more captious and critical do we become. There is many a good lady, who cannot tolerate a sewing-machine, although she knows it will do the work of ten seamstresses, because it will not sew on buttons and work buttonholes! Most of us are very much out of temper with the magnetic telegraph, just now, because it does not bring us the Court news from England every morning before breakfast, though we have hourly dispatches from Washington, New Orleans, and St. Louis; and, returning to our *moutons*, everybody is finding fault with us just now, because we cannot tell them of some universal, all-penetrating, cheap, strong, simple, enduring little implement, by means of which any kind of a laborer, Scotch, Irish, or Yankee, may conveniently open all kinds of drains in all kinds of land, whether sand, hard-pan, gravel, or clay. [226]

Having personally inquired and examined, touching draining tools in England, and having been solicited by an extensive agricultural implement house in Boston, to furnish them a list and description of a complete set of draining tools, and feeling the obligation which seemed to be imposed on us, to know all about this matter, we wrote to Mr. Denton, one of the first draining engineers in the world, to send us a list, with drawings and descriptions of such implements as he finds most useful, or, if more convenient the implements themselves.

237

Mr. Denton kindly replied to our inquiry, and his answer may be taken as the best evidence upon this point. He says:

"As to tools, it is the same with them as it is with the art of draining itself—too much rule and too much drawing upon paper; all very right to begin with, but very prejudicial to progress. I employ, as engineer to the General Land Drainage Company, and on my private account, during the drainage season, as many as 2,000 men, and it is an actual fact, that not one of them uses the set of tools figured in print. I have frequently purchased a number of sets of the Birmingham tools, and sent them down on extensive works. The laborers would purchase a few of the smaller tools, such as Nos. 290, 291, and 301, figured in Morton's excellent Cyclopædia of Agriculture, and would try them, and then order others of the country blacksmith, differing in several respects; less weighty and much less costly, and, moreover, much better as working tools. All I require of the cutters, is, that the bottom of the drain should be evenly cut, to fit the size of the pipe. The rest of the work takes care of itself; for a good workman will economize his labor for his own sake, by moving as little earth as practicable; thus, for instance, a first-class cutter, in clays, will get down four feet with a twelve-inch opening, *ordinarily*; if he wishes to *show off*, he will sacrifice his own comfort to appearance, and will do it with a ten-inch opening."

Having thus "freed our mind" by way of preliminary, we propose to take up our subject, and pursue it as practically and quietly as possible to the end. It may be well, perhaps, first to suggest by way of explanation of [227] Mr. Denton's letter, above quoted, that drains are usually opened in England by the yard, or rod, the laborer finding his own tools.

As has been intimated, the implements convenient for draining, depend on many circumstances. They depend upon the character of the earth to be moved. A sharp, light spade, which may work rapidly and well in a light loam or sand, may be entirely unfit to drive into a stiff clay; and the fancy bottoming tools which may cut out a soft clay or sand in nicely-measured slices, will be found quite too delicate for a hard-pan or gravel, where the pick-axe alone can open a passage.

The implements again must be suited to the workman who handles it. Henry Ward Beecher, in speaking of creeds, which he, on another occasion, had said were "the skins of religion set up and stuffed," remarked, that it was of more importance that a man should know how to make a practical use of his faith, than that he should subscribe to many articles; for, said he, "I have seen many a man who could do more at carpenter's work with one old jack-knife, than another could do with a whole chest of tools!"

What can an Irishman do with a chopping ax, and what cannot a Yankee do with it? Who ever saw a Scotchman or an Irishman who could not cut a straight ditch with a spade, and who ever saw a Yankee who could or would cut a ditch straight with any tool? One man works best with a long-handled spade, another prefers a short handle; one drives it into the earth with his right foot, another with his left. A laboring man, in general, works most easily with such tools as he is accustomed to handle; while theorizing implement-makers, working out their pattern by the light of reason, may produce such a tool as a man *ought* to work with, without adapting it at all to the capacity or taste of the laborer. A man should be measured for his tools, as much as for his garment, and not be [228] expected to fit himself to another's notions more than to another's coat.

If the land-owner proposes to act as his own engineer, the first instrument he will want to use is a Spirit Level, or some other contrivance by which he may ascertain the variations of the surface of his field. The natural way for a Yankee to get at the grades is to *guess* at them, and this, practically, is what is usually done. Ditches are opened where there appears to be a descent, and if there is water running, the rise is estimated by its current; and if there is no water rising in the drain, a bucketfull is occasionally poured in to guide the laborer in his work. No one who has not tested the accuracy, or, rather, inaccuracy, of his judgment, as to the levels of fields, can at all appreciate the deceitfulness of appearances on this point. The human eye will see straight; but it will not see level without a guide. It forms conclusions by comparison; and the lines of upland, of forest tops and of distant hills, all conspire to confuse the judgment, so that it is quite common for a brook to appear to the eye to run up hill, even when it has a quick current. A few trials with a spirit-level will cure any man of his conceit on this subject.

And so it is as to the regular inclination of the bottom of drains. It is desirable not only to have an inclination all the way, but a regular inclination, as nearly as possible, especially if the descent be small. Workmen are very apt to work at a uniform depth from the surface, and so give the bottom of the drain the same variations as the surface line; and thus at one point there may be a fall of one inch in a rod; at another, twice that fall; and at another, a dead level, or even a hollow. On our own farm, we have found, in twelve rods, a variation of a foot in the bottom line of a drain opened by skillful workmen on a nearly level field, where they had no water to guide them, and where they had supposed their fall was regular throughout. [229]

The following sketch shows the difference between lines of tiles laid with and without instruments. Next to guessing at the fall in our field, may be placed a little contrivance, of which we have made use sufficiently to become satisfied of its want of practical accuracy. It is thus figured and described in the excellent treatise of Thomas, on Farm Implements.

Fig. 54.

"*A* is a common square, placed in a slit in the top of the stake *B*. By means of a plumb-line the square is brought to a level, when a thumbscrew, at *C*, fixes it fast. If the square is two feet long, and is so carefully adjusted as not to vary more than the twentieth of an inch from a true level, which is easily accomplished, then a twentieth of an inch in two feet will be one inch in forty feet — a sufficient degree of accuracy for many cases."

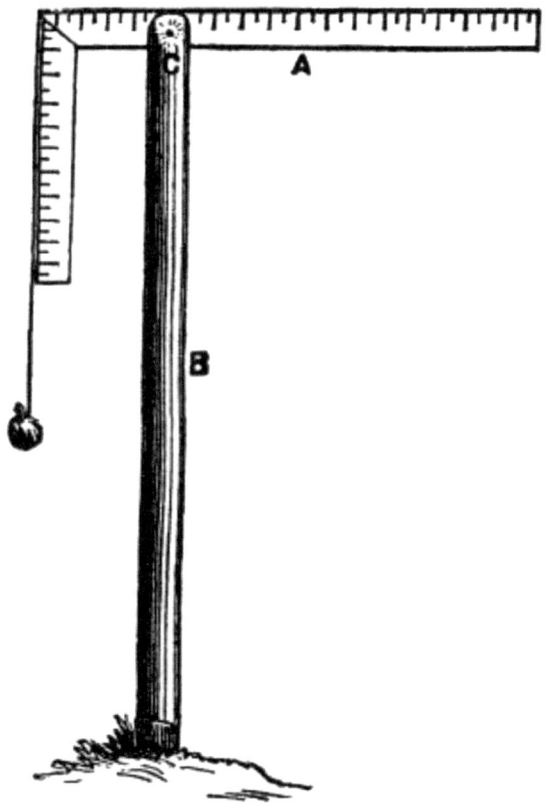

Fig. 55.—Square and Plumb-Level.

We do not so much object to the principle of the above level, as to its practical working. We find it difficult, without cross sights, to take an accurate level with any instrument. However, those who are used to rifle-shooting may hit tolerably near the mark with the square. Mr. Thomas only claims that it is accurate enough "for many cases."

A proper spirit-level, such as is used by engineers of railroads and canals, attached to a telescope, is the best of all instruments. "So great is the perfection of this [230] instrument," says the writer just quoted, "that separate lines of levels have been run with it, for sixty miles, without varying two-thirds of an inch for the whole distance." A cheap and convenient spirit-level, for our purpose, is thus constructed.

It is furnished with eye sights, *a b*, and, when in use, is placed into a framing of brass which operates as a spring to adjust it to the level position, *d*, by the action of the large-headed brass screw, *c*. A stud is affixed to the framing, and pushed firmly into a gimlet-hole in the top of the short rod, which is pushed or driven into the ground at the spot from whence the level is desired to be ascertained. It need scarcely be mentioned, that the height of the eye sight, from the guard, is to be deducted from the height of observation, which quantity is easily obtained by having the rod marked off in inches and feet; but it may be mentioned, that this instrument should be used in all cases of draining on level ground, even when one is confident that he knows the fall of the ground; for the eye is a very deceitful monitor for informing you of the levelness of ground. It is so light as to admit of being carried in the pocket, whilst its rod may be used as a staff or cane.

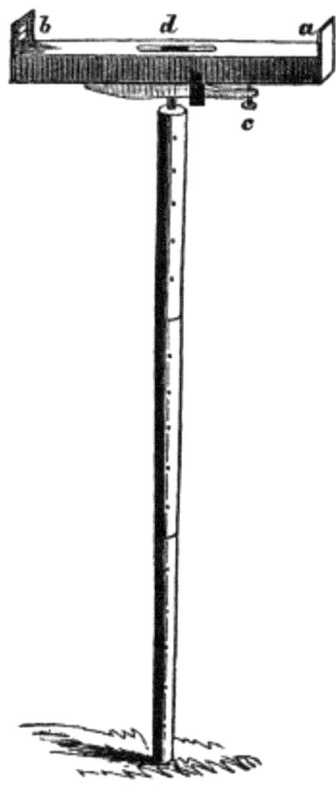

Fig. 56.—Spirit Level.

A staff of ten feet in length, graduated in feet and inches, and held by an attendant at the various points of observation, is necessary in the use of the spirit-level in the field. A painted target, arranged with a slide to be moved up and down on this staff, and held by a thumbscrew, will be found useful. [231]

We have made for our own use a level like the above, and find it sufficiently accurate for drainage purposes. Small spirit-levels set in iron can be had at the hardware shops for twenty cents each, and can be readily attached to wood by a screw, in constructing our implement; or a spirit-level set in mahogany, of suitable size, may be procured for a half dollar, and any person, handy with tools, can do the rest. The sights should be arranged both ways, with a slit cut with a chisel through the brass or tin, and an oblong opening at each end. The eye is placed at the slit, and sight is taken by a hair or fine thread, drawn across the opening at the other end. Then, by changing ends, and sighting through the other end at a given object, any error in the instrument may be detected. The hair or thread may be held in place by a little wax, and moved up or down till it is carefully adjusted. The instrument should turn upon the staff in all directions, so that the level of a whole field, so far as it is within range, may be taken from one position.

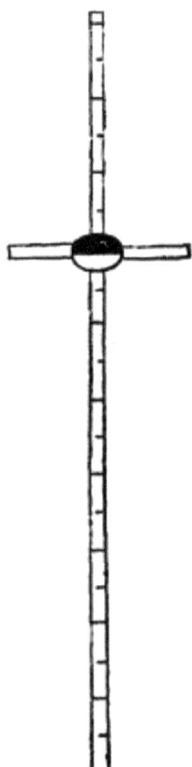

Fig. 57.
Staff and Target.

To maintain a uniform grade in the bottom of a drain so as to economize the fall, and distribute it equally through the whole length, several different instruments and means may be adopted. The first which we will figure, is what is called the Span, or A Level. Such a level may be easily constructed of common inch-board. If it be desired to note the fall in feet, the span may conveniently be ten feet. If a notation in rods be preferred, the span should be a rod, or half rod long.

The two feet being placed on a floor, and ascertained [232] to be perfectly level by a spirit-level, the plumb-line will hang in the centre, where a distinct mark should be made on the cross-bar. Then place a block of wood, exactly an inch thick, under one leg, and mark the place where the line crosses the bar. Put another block an inch thick under the same leg, and again mark where the line crosses the bar, and so on as far as is thought necessary. Then put the blocks under the other leg in the same manner, and mark the cross-bar. If the span be ten feet, the plumb-line will indicate upon the bar, by the mark which it crosses, the rise or fall in inches, in ten feet. If the span be a rod, the line will indicate the number of inches per rod of the rise or fall.

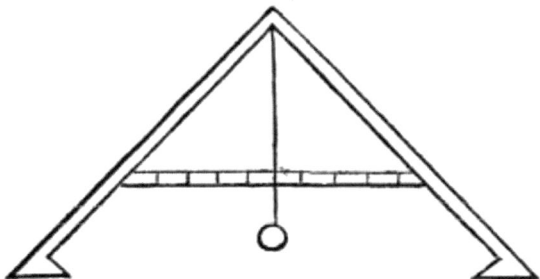

Fig. 58.—Span, or A Level.

This instrument is used thus: The fall of the ditch from end to end being ascertained by the spirit-level, and the length also, the fall per rod, or per one hundred feet, may be computed. The span is then placed in the bottom of the drain, from time to time, to guide the workman, or for accurate inspection of the finished cut. We have constructed and used this level, and found it very convenient to test the accuracy of the workmen, who had opened drains in our absence. A ten-foot span will be found as large as can be conveniently carried about the farm.

For the accurate grading of the bottom of drains, as the work proceeds, we have in practice found nothing so convenient and accurate as the arrangement which we are about to illustrate. [233]

The object is simply to draw a line parallel with the proposed bottom of the drain, for the laborers to work under, so that they, as they proceed, may measure down from it, as a guide to depth. Having with the spirit-level, ascertained the fall from end to end of the drain, a short stake is set at each end, and a line is drawn from one to the other at the requisite height, and supported by the cross-pieces, at suitable distances, to prevent the sagging of the line.

Fig. 59. — Grading Trenches by Lines.

Suppose the drain to be ten rods long, and that it is intended to cut it four feet deep, the natural fall being, from end to end, sufficient. We drive a stake at each end of the drain, high enough to attach to it a line three feet above the surface, which will be seven feet above the bottom of the finished drain — high enough to be above the heads of the cutters, when standing near the bottom.

Before drawing the line, the drain may be nearly completed. Then drive the intermediate stakes, with the projecting arms, which we will call squares, on one side of the drain, carefully sighting from one end of the stake to the other, at the point fixed for the line, and driving the squares till they are exactly even. Then attach a strong small cord, not larger than a chalk line, to one of the stakes, and draw it as tight as it will bear, and secure it [234] at the other stake. The line is now directly over the middle of the drain, seven feet from the bottom. Give the cutters, then, a rod seven feet long, and let them cut just deep enough for the rod to stand on the bottom and touch the line. Practically, this has been found by the author, the most accurate and satisfactory method of bringing drains to a regular grade.

Instead of a line, after the end stakes have been placed, a *boning rod*, as it is called, may be used thus: A staff is used, with a cross-piece at the top, and long enough, when resting on the proper bottom of the drain, to reach to the level of the marks on the stakes, three feet above the surface. Cross-pieces nailed to the stakes are the most conspicuous marks. A person stands at one stake sighting along to the other; a second person then holds the rod upright in the ditch, just touching the bottom, and carries it thus along. If, while it is moved along, its top is always in a line with the cross-bars on the end stakes, the fall is uniform; if it rise above, the bottom of the drain must be lowered; if it fall below, the bottom of the drain must be raised. This may be convenient enough for mere inspection of works, but it requires two persons besides the cutters, to finish the drain by this mode; whereas, with the lines and squares, any laborer can complete the work with exactness.

Another mode of levelling, by means of a mammoth mason's level, with an improvement, was invented by Colonel Challoner, and published in the Journal of the Royal Agricultural Society. It may appear to some persons more simple than the span level. We give the cut and explanation.

"I first ascertain what amount of fall I can obtain, from the head of every drain to my outfall. Suppose the length of the drain to be 96 yards, and I find I have a fall of two feet, that gives me a fall of a quarter of an inch in every yard. I take a common bricklayer's [235] level 12 feet long, to the bottom of which I attach, with screws, a piece of wood the whole length, *one inch wider* at one end than at the other, thereby throwing the level one inch out of the true horizontal line. When the drain has got to its proper depth at the outfall, I apply the broadest end of the level to the mouth; and when the plumb-bob indicates the level to be correct, the one-inch fall has been gained in the four yards, and so on. I keep testing the drain as it is dug, quite up to the head, when an unbroken, even, and continuous fall of two feet in the whole 96 yards has been obtained."

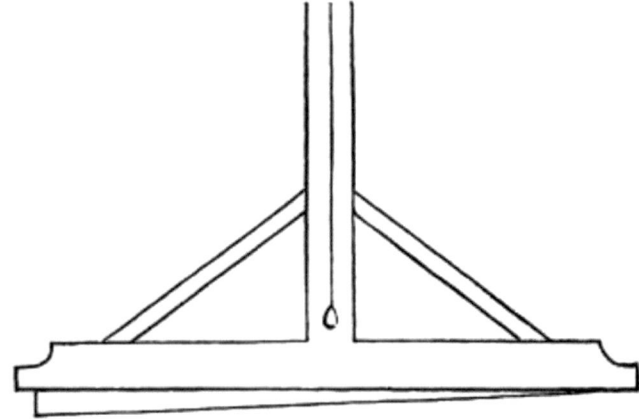

Fig. 60. — Challoner's Level.

## SPADES AND SHOVELS.

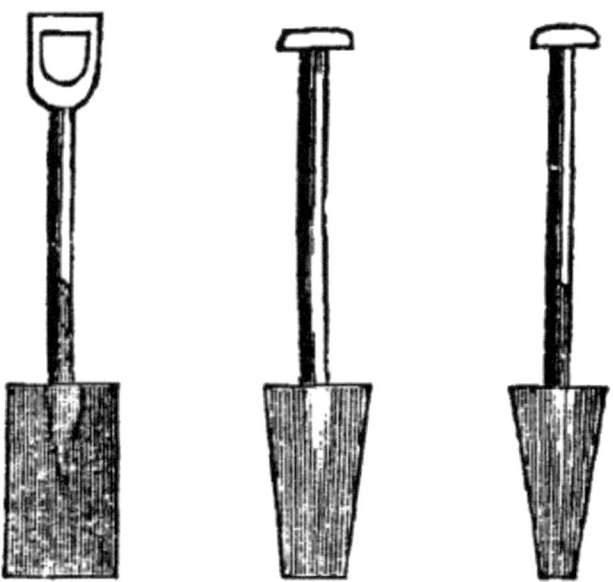

Fig. 61, 62, 63. — Drain Spades.

No peculiar tool is essential in opening that part of the drain which is more than a foot in width. Shovels and spades, of the forms usually found upon well-furnished farms, and adapted to its soil, will be found sufficient. A Boston agricultural house, a year or two since, sent out an order to London for a complete set of draining tools. In due season, they received, in compliance with their order, three spades of different width, like those represented in the cut.

These are understood to be the tools in common use in [236] England and Scotland, for sod-draining, and for any other drains, indeed, except tiles. The widest is 12 inches wide, and is used to remove the first spit, of about one foot depth. The second is 12 inches wide at top, and 8 at the point, and the third, eight at top, and four at the point. The narrowest spade is usually made with a spur in front, or what the Irish call a *treader*, on which to place the foot in driving it into the earth.

Fig. 64.
Spade with Spur.

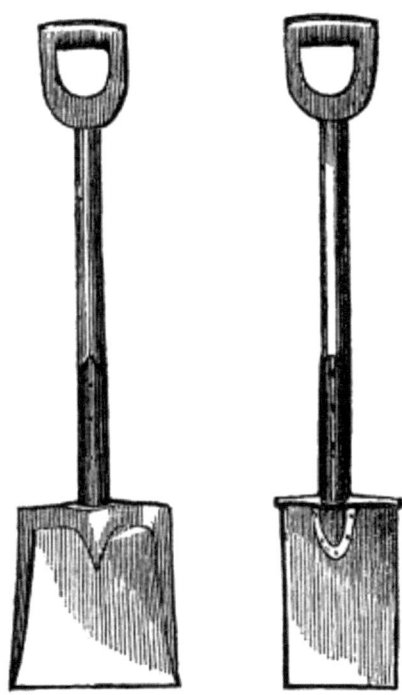

Figs. 65, 66. — Common Shovel and Spade.

For wedge drains, these spades are made narrower than those above represented, the finishing spade being but two and a half inches wide at the point. It will be recollected that this kind of drainage is only adapted to clay land. The shovels and spades which have been heretofore in most common use in New England are made with short handles, thus —

They are of cast-steel, and combine great strength and lightness. Long-handled shovels and spades are much preferred, usually, by Irish laborers, whose fancy is worth consulting in matters with which they have so much to do. We believe their notion is correct, that the long-handled tool is the easier to work with, at almost any job.

In our own draining, we find the common spade, with long or short handle, to be best in marking out the lines in turf; and either the spade or common shovel, according [237] to the nature of the soil, most convenient in removing the first foot of earth.

After this, if the pick is used, a long-handled round-pointed shovel, now in common use on our farms, is found convenient, until the ditch is too narrow for its use. Then the same shovel, turned up at the sides so as to form a narrow scoop, will be found better than any tool we yet have to remove this loosened earth.

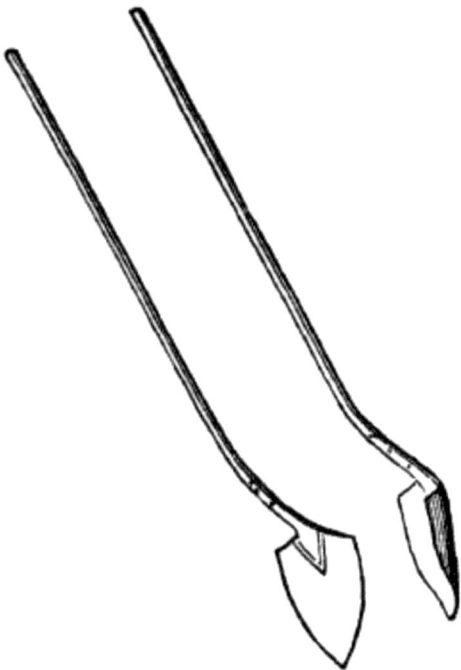

Figs. 67, 68. — Long-handled Round Shovel. Scoop Shovel.

Of all the tools that we have ever seen in the hands of an Irishman, in ditching, nothing approximates to the true Irish spade. It is a very clumsy, ungainly-looking implement used in the old country both for ditching, and for ridging for potatoes, being varied somewhat in width, according to the intended use. For stony soil, it is made narrower and stronger, while for the bog it is broader and lighter. The Irish blacksmiths in this country usually know how to make them, and we have got up a pattern of them, which are manufactured by Laighton and Lufkin, edge-tool makers, of Auburn, N. H., which have been tested, and found to suit the ideas of the Irish workmen.

This is a correct portrait of an Irish spade of our own pattern, which has done more in opening two miles of drains on our own farm, than any other implement.

The spade of the Laighton and Lufkin pattern weighs 5 lbs., without the handle, and is eighteen inches long. It is of iron, except about eight inches of the blade, which is of cast steel, tempered and polished like a chopping axe. It is considerably curved, and the workmen suit their own taste as to the degree of curvature, by putting [238] the tool under a log or rock, and bending it to suit themselves. It is a powerful, strong implement, and will cut off a root of an inch or two diameter as readily as an axe. The handle is of tough ash, and held in place by a wedge driven at the side of it, and can be knocked out readily when the spade needs new steel, or any repair. The length of the handle is three feet eight inches, and the diameter about one and one-fourth inches. The wedge projects, and forms a "treader," broad and firm, on which the foot comes down, to drive the spade into the ground.

Fig. 69.
Irish Spade.

We have endeavored to have the market supplied with the Irish spades, because, in the hands of such Irishmen as have used them "at home," we find them a most effective tool. We are met with all sorts of reasonable theoretical objections on the part of implement sellers, and of farmers, who never saw an Irish spade in use. "Would not the tool be better if it were wider and lighter," asks one. "I think it would be better if the spur, or "treader," were movable and of iron, so as to be put on the other side or in front," suggests another. "It seems as if it would work better, if it were straight," adds a third. "Would it not hold the dirt better if it were a little hollowing on the front," queries a fourth. "No doubt," we reply, "there might be a very good implement made, wider and lighter, without a wooden treader, and turned up at the sides, to hold the earth better, but it would not be an Irish spade when finished. Your theories may

254

be all correct and demonstrable by the purest mathematics, but the question is, with what tool will Patrick do the most work? If he [239] recognizes the Irish spade as an institution of his country, as a part of 'home,' you might as well attempt to reason him out of his faith in the Pope, as convince him that his spade is not perfect." Our man, James, believes in the infallibility of both. There is no digging on the farm that his spade is not adapted to. To mark out a drain in the turf by a line, he mounts his spade with one foot, and hops backward on the other, with a celerity surprising to behold. Then he cuts the sod in squares, and, with a sleight of hand, which does not come by nature, as Dogberry says reading and writing come, throws out the first spit. When he comes on to the gravel or hard clay, where another man would use a pick-axe, his heavy boot comes down upon the treader, and drives the spade a foot or more deep; and if a root is encountered, a blow or two easily severs it. The last foot at the bottom of the four-foot drain, is cut out for the sole-tile only four and a half inches wide, and the sides of the ditch are kept trimmed, even and straight, with the sharp steel edge. And it is pleasant to hear James express his satisfaction with his national implement. "And, sure, we could do nothing at this job, sir, without the Irish spade!" "And, sure, I should like to see a man that will spade this hard clay with anything else, sir!" On the whole, though the Irish spade does wonders on our farm, we recommend it only for Irishmen, who know how to handle it. In our own hands, it is as awkward a thing as we ever took hold of, and we never saw any man but an Irishman, who could use it gracefully and effectively.

*Bottoming Tools.*—The only tools which are wanted of peculiar form in draining, are such as are used in forming the narrow part of the trenches at the bottom. We can get down two feet, or even three, with the common spade and pick-axe, and in most kinds of drainage, except with tiles, it is necessary to have the bottom as wide, at least, [240] as a spade. In tile-draining, the narrower the trench the better, and in laying cylindrical pipes without collars, the bottom of the drain should exactly fit the pipes, to hold them in line.

Although round pipes are generally used in England, we have known none used in America until the past season—the sole-pipe taking their place. As the sole-pipe has a flat bottom, a different tool is required to finish its resting-place, from that adapted to the round pipe. As we have not, however, arrived quite at the bottom, we will return to the tools for removing the last foot of earth.

And first, we give from Morton, the Birmingham spades referred to by Mr. Denton, in his letter, quoted in this chapter. They are the theoretically perfect tools for removing the last eighteen or twenty inches of soil in a four or five-foot drain. Mr. Gisborne says of the drain properly formed:

"It is wrought in the shape of a wedge, brought in the bottom to the narrowest limit which will admit the collar, by tools admirably adapted to that purpose. The foot of the operator is never within twenty inches of the floor of the drain; his tools are made of iron, plated on steel, and never lose their sharpness, even when worn to the stumps; because, as the softer material, the iron, wears away, the sharp steel edge is always prominent."

Fig. 70.

Fig. 71.

Fig. 72.

Birmingham Spades.

This poetical view of digging drains, meets us at every turn, and we are beset with inquiries for [241] these wonderful implements. We do not intimate that Mr. Gisborne, and those who so often quote the above language, are not reliable. Mr. Gisborne "is an honorable man, so are they all honorable men;" but we must reform our tiles, and our land too, most of it, we fear, before we can open four-foot trenches, and lay pipes in them, without putting a foot "within twenty inches of the floor of the drain."

In the first place, we have great doubt whether pipes can be laid close enough to make the joints secure without collars, unless carefully laid by hand, or unless they are round pipes, rolled in the making, when half dried, and so made straight and even at the ends. In laying such sole-pipes as we have laid, it requires some care to adjust them, so as to make the joints close. Most of them are warped in drying or burning, so that spaces of half an inch will often be left at the top or side, where two are laid end to end. Now, if the foot never goes to the bottom of the drain, the pipes must be laid with a hook or pipe-layer, such as will be presently described, which may do well for pipes and collars, because the collar covers the joint, so that it is of no importance if it be somewhat open.

Again, we know of no method of working with a pick-axe, except by standing as low as the bottom of the work. No man can pick twenty inches, or indeed any inches, lower than he stands, because he must move forward in this work, and not backward. Each land-owner may judge for himself, whether his land requires the pick in its excavation.

In soft clays, no doubt, with suitable tools, the trench may be cut a foot, or more, lower than the feet of the workman. We have seen it done in our land, in a sandy soil, with the Irish spade, though, as we used sole-pipes, our "pipe-layer" was a live Irishman, who walked in the [242] trench backwards, putting down the pipes with his hand.

We are satisfied, that the instances in which trenches may be opened a foot or two below the feet of the workmen, are the exceptions, and not the rule, and that in laying sole-tiles, the hand of a careful workman must adjust each tile in its position.

We have found a narrow spade, four inches wide, with a long handle, a convenient tool for finishing drains for sole-tiles.

Fig. 73.

Fig. 74.

Narrow Spades for Tiles.

We have thoroughly tested the matter; and in all kinds of soil, give a decided preference to spades as broad at the point as at the heel. We have used common long-handled spades, cut down with shears at a machine-shop, into these shapes.

The spade of equal width, works much more easily in the bottom of a trench, because its corners do not catch, as do those of the other. The pointed spade is apparently nearer the shape of the sloping ditch, but such tools cannot be used vertically, and when the heel of the pointed spade is lowered, it catches in the side of the trench, before the point reaches the bottom.

Very strong spades, of various width, from three to eight inches, and thick at the heel, to operate as a wedge, will be found most suitable for common use. The narrowest spades should have the spur, as shown in Fig. [243] 64, because there is not room for the foot by the side of the handle.

The various tools for finishing the bottoms of drains, as figured in Morton, are the following:

Fig. 75.

Fig. 76.

Fig. 77.

Fig. 78.

Fig. 79.

English Bottoming Tools.

The last implement, which is a scoop for the bottom of trenches for round pipes, is one of the tools mentioned in Mr. Denton's letter, as not being found to the taste of his workmen. For scooping out our flat-bottomed trenches, we use a tool like Fig. 77. For boggy land, soft clay, or, indeed, any land where water is running at the time of [244] the excavation, scoops like the following will be found convenient for flat bottoms.

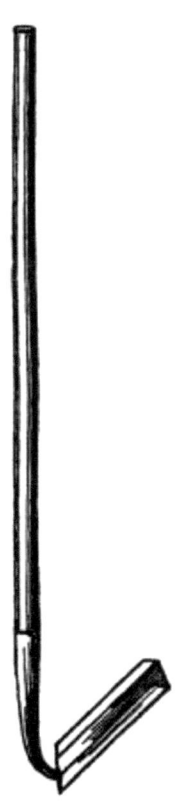

Fig. 80.

Fig. 81.

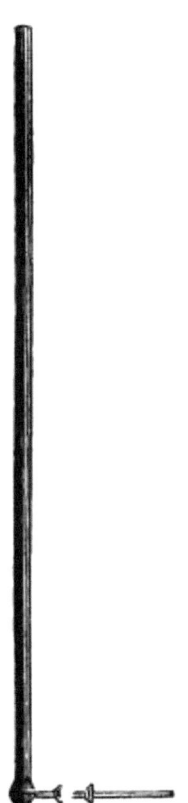

Fig. 82.

Drawing and Pushing Scoop, and Pipe-Layer.

The pushing scoop (Fig. 81), as it is called, may be made of a common long-handled shovel, turned up at the sides by a blacksmith, leaving it of the desired width.

The *pipe-layer*, of which mention has so often been made, is a little implement invented by Mr. Parkes, for placing round pipes and collars in narrow trenches, without stepping into them.

The following sketch, by our friend Mr. Shedd, shows the pipe-layer in use. The cross section of the land, shown in front, represents it as having had the advantage of draining, by which the water-table is brought to a level with the bottom of the drain, as shown by the heavy shading. [245] An "Irish spade" and a pipe-layer are shown lying on the ground.

Fig. 83. — Pipe-Laying.

The *pick-axes* commonly used in excavation of trenches, are in the following forms:

[246]

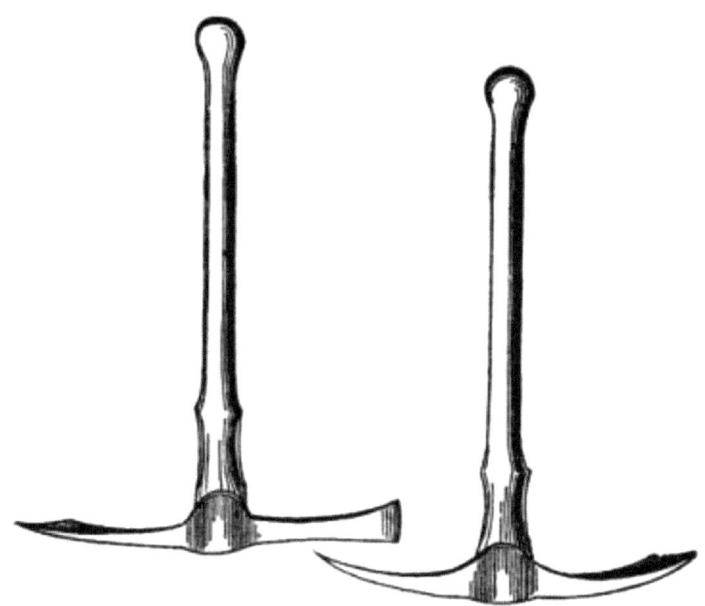

Fig. 84, 85.—Pick-axes.

Pick-axes may be light or heavy, according to the nature of the soil. A chisel at one end, and point at the other, is found best in most cases.

A *Drain-gauge* is usually mentioned in a list of draining tools. It is used when ditches are designed for stone or other material than tiles, and where the width is important. In tile-draining the width is entirely immaterial. If opened by the rod, it is only important that they be of proper depth and inclination, with the bottom wide enough for the tile.

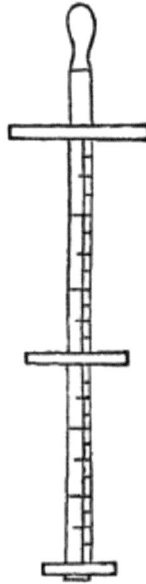

Fig. 86. — Drain-Gauge.

The above figure shows the usual form of the drain-gauge. Below, we give from Morton, drawings, and a description of Elkington's augers for boring in the bottoms of ditches.

"The cut annexed represents the auger employed by Elkington, where *a b* and *c* are different forms of the tool; *d*, a portion of the shaft: *e*, with the wedges, *h h*, the cross handle; and *f* and *g* additional pieces for grasping the shaft, and so enabling more than one person to work at it." The auger-hole ought to be a little at one side of the drain, as in Fig. 3, at page 35, so that the water may not rise at right angles to the flow of water in it, and obstruct its current.

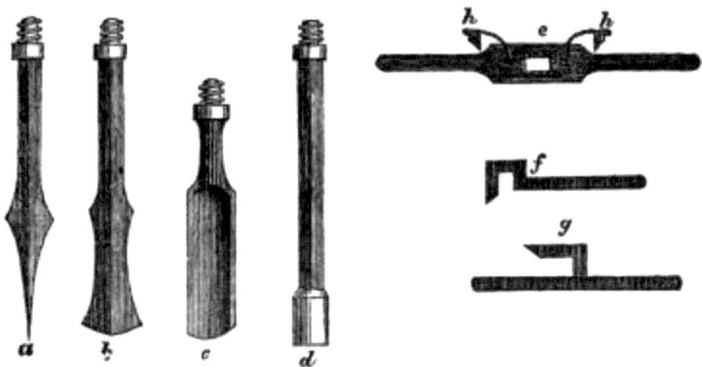

Fig. 87.—Elkington's Draining Auger.

- *a.* The plug, or point under ground, to which the string of pipes is attached.
- *bb.* The coulter from the point up through the beam, regulated by wheel and screw midway.
- *c.* The beam connecting the two pairs of wheels.
- *e.* Drain opened by hand where pipes enter the ground.
- *a to e.* Pipes under ground.
- *e to f.* Pipes above ground.
- *g.* Windlass or capstan, worked by horses.
- *h.* Wire rope attached to plow, and wound round the windlass.

- *i.* Pulley round which the rope runs to keep the plow in the line of the ditch.

## DRAINING-PLOWS AND DITCH-DIGGERS. [247]

The man who can invent and construct a machine that shall be capable of cutting four-foot ditches for pipe-drains, with facility, will deserve well of his country.

It is not essential that the drain be cut to its full depth at one operation. If worked by oxen or horses, it may go several times over the work, taking out a few inches at each time. If moved by a capstan, or other slowly-operating power, it must work more thoroughly, so as not to consume too much time.

With a lever, such as is used in Willis's Stump Puller, sufficient power for any purpose may be applied. An implement like a subsoil plow, constructed to run four feet deep, and merely doing the work of the pick, would be of great assistance. Prof. Mapes says he has made use of such an implement with great advantage. For tile-drains, the narrower the ditch the better, if it be only wide enough to receive the tiles. A mere slit, four inches wide, if straight and of even inclination at the bottom, would be the best kind of ditch, the pipes being laid in with a pipe-layer. But if the ditch is to be finished by the machine, it is essential that it be so contrived that it will grade the bottom, and not leave it undulating like the surface. Fowler's Drain Plow is said to be so arranged, by improvements since its first trials, as to attain this object.

Having thus briefly suggested some of the points to be kept in mind by inventors, we will proceed to give some account of such machines as come nearest to the wants of the community. Fowler's Draining-Plow would meet the largest wants of the public, were it cheap enough, and really reliable to perform what it is said to perform. The author saw this implement in England, but not in operation, and it seems impossible, from inspection of it, [248] as well as from the theory of its operation, that it can succeed, if at all, in any but soft homogeneous clay. The idea is, however, so bold, and so much is claimed for the implement, that some description of it seems indispensable in a work like this.

The pipes, of common drain tiles, are strung on a rope, and this rope, with the pipes, is drawn through the ground, following a plug like the foot of a subsoil plow, leaving the pipes perfectly laid, and the drain completed at a single operation. (See Fig. 88.)

The work is commenced by opening a short piece of ditch by hand, and strings of pipes, each about 50 feet long, are added as the work proceeds; and when the ditch is completed, the rope is withdrawn. When the surface is uneven, the uniform slope is preserved by means of a wheel and screw, which governs the plug, or coulter, raising or lowering it at pleasure. A man upon the frame-work controls this wheel, guided by a sight on the frame, and a cross-staff at the end of the field.

Drains, 40 rods long, are finished at one operation. This plow has been carefully tested in England. Its work has been uncovered when completed, and found perfect in every respect. The great expense of the machine, and the fact that it is only adapted to clay land free from obstructions, has prevented its general use. We cannot help believing that, by the aid of steam, on our prairies, at least, some such machine may be found practicable and economical.

## PRATT'S DITCH DIGGER,

Patented by Pratt & Bro., of Canandaigua, is attracting much attention. We have not seen it in operation, nor have we seen statements which satisfy us that it is just what is demanded. It is stated, in the *Country Gentleman*, to be incapable of cutting a ditch more than two and a half feet deep. A machine that will do so much is [249] not to be despised; but more than one half the digging remains of a four-foot ditch, after two and a half feet are opened, and we want an implement to do the lowest and worst half. It is stated that, in one instance, a ditch, 60 rods long, about two feet deep, in hard clay, was cut with this machine, worked by two horses, in five hours.

We trust that the enterprising inventors will perfect their implement, so that it will open drains four feet deep, and thus meet the great want of the public. It is not to be expected that any such implement can be made to operate in ground full of stones and roots; and inventors should not be discouraged by the continual croakings of those sinister birds, which see nothing but obstacles, and prophecy only failure.

Fig. 89.—Pratt's Ditch Digger.

The drain plow was first introduced into Scotland by M'Ewan. The soil in his district was mostly a strong unctuous clay, free from stones. He constructed an immense plow, worked by 12 or 16 horses, by means of which a furrow-slice, 16 inches in depth, was turned out; and, by a modification of the plow, a second slice was removed, to the depth, in all, of two feet. This plow is expensive and heavy, and incapable of working to sufficient depth.

Mr. Paul, of Norfolk Co., England, has lately invented an ingenious machine for cutting drains, of which we give an elevation.

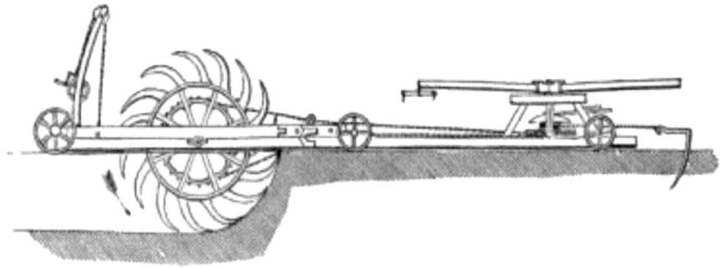

Fig. 90.—Paul's Ditching Machine.

It is worked by a chain and capstan, by horses, and, of [250] course, may be operated by steam or lever power. It is drawn forward, and, as it moves, it acts as a slotting machine on the land, the tools on the circumference of the acting-wheel taking successive bites of the soil, each lifting a portion from the full depth to which it is desired that the trench should be cut, and laying the earth thus removed on the surface at either side. There is a lifting apparatus at the end of the machine, by which the cutting-wheel may be raised or lowered, according to the unevenness of the surface, in order to secure a uniform fall in the bottom of the drain. The whole process is carried on at the rate of about four feet per minute, and it results, on suitable soils, in cutting a drain from three to five feet deep, leaving it in a finished state, with a level bottom for the tiles to rest upon. We give the cut and statement from the Cyclopædia of Agriculture, and if the machine shall prove what it is represented to be, we see but little more to be desired in a ditching machine. The principle of this implement appears to us to be the correct one, [251] and we see no reason to doubt the statement of its performance.

Routt's drain plow is designed for surface-draining merely. We give, from the *New England Farmer*, a statement of its merits, as detailed by a correspondent who saw it at the exhibition of the U. S. Agricultural Society at Richmond, in 1858:

"One of the most attractive implements on the Fair ground, to the farmer, was A. P. Routt's patent drain plow. This implement makes a furrow a foot deep, two feet and a half wide at the top, and four inches wide at the bottom, the sides sloping at such an angle as to insure the drain from falling in by the frost, the whole being perfectly completed at one operation by this plow, or tool. Those who have tried it say it is the very thing for surface-draining, which, on wet lands, is certainly very beneficial where under-draining has not been done. The manufacturer resides in Somerset, Orange County, Va. The plow is so made that it opens a deep furrow, turning both to the right and left, and is followed by a heavy iron roller that hardens the earth, both on the sides and the bottom of the surface-drain, thus doing very handsome work. The price, as heretofore stated, is $25, and with it, a man can, with a good pair of team horses, surface-drain 60 acres of land a day."

# CHAPTER XII [252]
# PRACTICAL DIRECTIONS FOR OPENING DRAINS AND LAYING TILES.

Begin at the Outlet. — Use of Plows. — Levelling the Bottom. — Where to begin to lay Pipes. — Mode of Procedure. — Covering Pipes. — Securing Joints. — Filling. — Securing Outlets. — Plans.

In former chapters, we have spoken minutely of the arrangement, depth, distance, and width of drains; and in treating of tools for drainage, we have sufficiently described the use of levelling instruments and of the various digging tools.

We assume here, that the engineering has been already done, and that the whole system has been carefully staked out, so that every main, sub-main, and minor is distinctly located, and the fall accurately ascertained. Until so much has been accomplished, we are unprepared to put the first spade into the ground.

We propose to give our own experience as to the convenient method of procedure, with such suggestions as occur to us, for those who are differently situated from ourselves.

The work of excavation must begin at the outlet, so that whatever water is met with, may pass readily away; and the outlet must be kept always low enough for this purpose. If there is considerable fall, it may not be best to deepen the lower end of the main to its full extent, at first, because the main, though first opened, must be the [253] last in which the pipes are laid, and may cave in, if unnecessarily deep at first. In many cases there is fall enough, so that the upper minors may be laid and find sufficient fall, before the lower end of the main is half opened.

With a garden line drawn straight, mark out the drain, with a sharp spade, on both sides, and remove the turf. If it is desired to use the turf for covering the pipes, or to replace it over the drains, when finished, it should at first be placed in heaps outside the line of the earth to be thrown out.

A plow is used sometimes to turn out the sod and soil; but we have few plowmen who can go straight enough; and in plowing, the soil is left too near to the ditch for convenience, and the turf is torn in pieces and buried, so as not to be fit for use. Usually, it will be found convenient to remove the turf, if there be any, with a spade, by a line. Then, a plow may be used for turning out the next spit, and the drain may be kept straight, which is indispensable to good work. A good ditching-machine is, of course, the thing needful; but we are endeavoring in these directions to do our best without it. We have opened our own trenches entirely by hand labor, finding laborers more convenient than oxen or horses, and no more expensive.

Many have used the plow in the first foot or two of the cutting, but it is not here "the first step which costs," but the later steps. After the first foot is removed, if the ground be hard, a pick or subsoil plow must be used. A subsoil plow, properly constructed, may be made very useful in breaking up the subsoil, though there is a difficulty in working cattle astride of a deep ditch, encumbered with banks of earth. A friend of ours used, in opening drains, a large bull in single harness, trained to walk in the ditch; but the width of a big bull is a somewhat [254] larger pattern for a drain, than will be found economical.

The ingenuity of farmers in the use of a pair of heavy wheels, with a chain attached to the axle, so that the cattle may both walk on one side of the ditch, or by the use of long double-trees, so that horses may go outside the banks of earth, will generally be found sufficient to make the most of their means.

It will be found convenient to place the soil at one side, and the subsoil at the other, for convenience in returning both right side up to their places.

Having worked down to the depth of two feet or more, the ditch should be too narrow for the use of common spades, and the narrow tools already described will be found useful. The Irish spade, on our own fields, is in use from the first to the last of the excavation; and at three feet depth, we have our trench but about six inches in width, and at the bottom, at four feet depth, it is but four inches—just wide enough for the laborer to stand in it, with one foot before the other.

Having excavated to nearly our depth, we use the lines, as described in another place, for levelling, and the men working under them, grade the bottom as accurately as possible. If flat-bottomed tiles are used, the ditch is ready for them. If round pipes are used, a round bottoming tool must be used to form a semi-circular groove in which the pipes are to lie.

We have not forgotten that English drainers tell us of tools and their use, whereby drains may be open twenty inches lower than the feet of the workman; but we have never chanced to see that operation, and are skeptical as to the fact that work can thus be performed economically, except in very peculiar soils. That such a *crack* may be thus opened, is not doubted; but we conceive of no means by which earth, that requires the pick, can be moved to [255] advantage, without the workman standing as low as his work.

Having opened the main, and finished, as we have described, the minor which enters the main at its highest point, we are ready to lay the tiles.

By first laying the upper drain, it will be seen that we may finish and secure our work to the junction of the first minor with its main.

Convey the pipes by wagon or otherwise, as is convenient, to the side of the ditch where the soil lies, and where there is least earth, and lay them close to the edge of the ditch, end to end the whole way, discarding all imperfect pieces. If it is designed to use gravel, turf, or other covering for the pipes, lay it also in heaps along the trench. Then place the first pipe at the upper end of the ditch, with a brick or stone against its upper end, to exclude earth. We have heretofore used sole-tiles, with flat bottoms, and have found that a thin chip of wood, not an eighth of an inch thick, and four by two inches in size, such as may be found at shoe shops in New England, assists very much in securing an even bearing for the tiles. It is placed so that the ends of two tiles rest on it, and serves to keep them in line till secured by the earth. A man walking backward in the ditch, takes the tiles from the bank, carefully adjusting them in line and so as to make good joints, and he can lay half a mile or more in a day, if the bottom is well graded. Another should follow on the bank, throwing in a shovel full of gravel or tan, if either is used, upon the joint.

If turf is to be used to secure the joint, pieces should be cut thin and narrow, and laid along the bank, and the man in the ditch must secure each joint as he proceeds. It will be found to cost twice the labor, at least, to use turf, as it is to use gravel or tan, if they are at hand.

If the soil be clay, we do not believe it is best to return [256] it directly upon the tiles, because it is liable to puddle and stop the joint, and then to crack and admit silt at the joint, while gravel is not thus affected. We prefer to place the top soil of clay land, next the pipes, rather than the clay in the condition in which it is usually found.

As to small stones above the pipes, we should decidedly object to them. They are unnecessary to the operation of the drain, and they allow the water to come in, in currents, on to the top of the pipes, in heavy storms or showers, and so endanger their security. The practice of placing stones above the tiles is abandoned by all scientific drainers.

We have, in England, seen straw placed over the joints of pipes, but it seems an inconvenient and insecure practice. Long straw cannot be well placed in such narrow openings, and it is likely to sustain the earth enough, so that when thrown in, it will not settle equally around the pipes; whereas a shovelfull of gravel or other earth sifted in carefully, will at once fasten them in place.

Having laid and partially covered the first or upper drain, proceed with the next in the same way, laying and securing the main or sub-main, at the same time, to each intersection, thus carrying the work from the highest point down towards the outlet. After sufficient earth has been thrown in to make the work safe against accidents by rain, or caving in of earth, the filling may be completed at leisure. Mr. Johnston, of Geneva, uses for this purpose a plow, having a double-tree nine and a half feet long, to enable a horse to go on each side of the ditch.

We suggest that a side-hill plow might well enough be used with horses *tandem*, or with oxen and cart wheels and draughts.

The filling, however, will be found a small matter, compared with the digging. In laying pipes in narrow trenches, a tool called a pipe-layer is sometimes used, a [257] cut of which, showing its mode of use, may be found in another place.

In filling drains where the soil is partly clay, and partly sand or gravel, we recommend that the clay be placed in the upper part of the drain, so as to prevent water from passing directly down upon the pipes, by which they are frequently displaced as soon as laid.

If the work is completed in Autumn, it is well to turn two or three furrows from each side on to the drains, so as to raise the surface there, and prevent water from cutting out the ditch, or standing above it. If the land is plowed in Autumn, it is best to back-furrow on to the drains, leaving dead furrows half way between them, the first season.

As to the importance of securing the outlets, and the manner of doing it, we have spoken particularly elsewhere.

And here, again, we will remind the beginner, of the necessity of making and preserving accurate plans of the work, so that every drain may be at any time found by measurement. After a single rotation, it is frequently utterly impossible to perceive upon the surface any indication of the line of the drains.

In this connection, it may be well perhaps to remind the reader, that whatever arrangements are made as to silt-basins, or peep-holes, must be included in the general plan, and executed as the work proceeds.

# CHAPTER XIII [258]
# EFFECTS OF DRAINAGE UPON THE CONDI-
# TION OF THE SOIL.

Drainage deepens the Soil, and gives the roots a larger pasture. — Cobbett's Lucerne 30 feet deep. — Mechi's Parsnips 13 feet long! — Drainage promotes Pulverization. — Prevents Surface-Washing. — Lengthens the Season. — Prevents Freezing out. — Dispenses with Open Ditches. — Saves 25 per cent. of Labor. — Promotes absorption of Fertilizing Substances from the Air. — Supplies Air to the Roots. — Drains run before Rain; so do some Springs. — Drainage warms the Soil. — Corn sprouts at 55°; Rye on Ice. — Cold from Evaporation. — Heat will not pass downward in Water. — Count Rumford's Experiments with Hot Water on Ice. — Aeration of Soil by Drains.

The benefits which high-lands, as we ordinarily call them, in distinction from swamp or flowed lands, derive from drainage, may be arranged in two classes, *mechanical* and *chemical*; though it is not easy, nor, indeed, is it important, to maintain this distinction in all points. Among those which partake rather of the nature of mechanical changes, are the following:

*Drainage deepens the soil.* Every one who has attempted to raise deep-rooted vegetables upon half-drained swamp-land, has observed the utter impossibility of inducing them to extend downward their usual length. Parsnips and carrots, on such land, frequently grow large at the top, but divide into numerous small fibres just below the surface, and spread in all directions. No root, except those of aquatic plants, will grow in stagnant water. If, therefore, it is of any advantage to have a deep, rather than a shallow soil, it is manifestly necessary, from this consideration alone, to lower the line of standing water, [259] at least, to the extent to which the roots of our cultivated crops descend. A deep soil is better than a shallow one, because it furnishes a more extensive feeding-ground for the roots. The elements of nutrition, which the plant finds in the soil, are not all upon the surface. Many of them are washed down by the rains into the subsoil, and some are found in the decomposing rocks themselves. These, the plants, by a sort of instinct, search out and find, as well in the depths of the earth as at its surface, if no obstacle opposes. By striking deep roots again, the plants stand more firmly in the earth, so that they are not so readily drawn out, or shaken by the winds. Indeed, every one knows that a soil two feet deep is better than one a foot deep; and market-gardeners and nursery-men show, by their practice, that they know, if others do not, that a trenched soil three feet deep is better than one of any less depth. We all know that Indian corn, in a dry soil, sends down its rootlets two feet or more, as well as most of the grasses. Cobbett says: "The lucerne will send its roots thirty feet into a dry bottom!" The Chinese yam, recently introduced, grows downward two or three feet. The digging of an acre of such a crop, by the way, on New England soil generally, would require a corps of sappers and miners, especially when we consider that the yam grows largest end downward. However, the yam may prove a valuable acquisition to the country. Every inch of additional soil gives 100 tons of active soil per acre.

Says Mr. Denton:

"I have evidence now before me, that the roots of the wheat plant, the mangold wurzel, the cabbage, and the white turnip, frequently descend into the soil to the depth of three feet. I have myself traced the roots of wheat nine feet deep. I have discovered the roots of perennial grasses in drains four feet deep; and I may refer to Mr. Mercer, of Newton, in Lancashire, who has traced the roots of rye grass running for many feet along a small pipe-drain, after descending four feet through the soil. Mr. Hetley, of Orton, assures me that he [260] discovered the roots of the mangolds, in a recently made drain, five feet deep; and the late Sir John Conroy had many newly-made drains, four feet deep, stopped by the roots of the same plants."

Mr. Sheriff Mechi's parsnips, however, distance anything in the way of deep rooting that has yet been recorded. The Sheriff is a very deep drainer, and an enthusiast in agriculture, and Nature seems to delight to humor his tastes, by performing a great many experiments at his famous place called Tiptree Hall. He stated, at a public meeting, that, in his neighborhood, where a crop of parsnips was growing on the edge of a clay pit, the roots were observed to descend 13 feet 6 inches; in fact, the whole depth to which this pit had once been filled up!

*Drainage assists pulverization.* It was Tull's theory that, by the comminution, or minute division, of soils alone, without the application of any manures, their fertility might be permanently maintained; and he so far supported this theory as, by repeated plowings, to produce twelve successive crops of wheat on the same land, without manure. The theory has received support from the known fact, that most soils are benefitted by Summer fallowing. The experiments instituted for the purpose of establishing this theory, although they disproved it, showed the great value of thorough pulverization. It is manifest that a wet soil can never be pulverized. Plowing clayey, or even loamy soil, when wet, tends rather to press it together, and render it less pervious to air and water.

The first effect of under-draining is to dry the surface-soil, to draw out all the water that will run out of it, so that, in early Spring, or in Autumn, it may be worked with the plow as advantageously as undrained lands in mid-Summer.

Striking illustrations of the benefits of thorough pulverization will be found in the excellent remarks of Dr. Madden, given in a subsequent chapter. [261]

*Drainage prevents surface-washing.* All land which is not level, and is not in grass, is liable to great loss by heavy rains in Spring and Autumn. If the land is already filled with water, or has not sufficient drainage, the rain cannot pass directly downward, but runs away upon the surface, carrying with it much of the soil, and washing out of what remains, of the valuable elements of fertility which have been applied with such expense. If the land be properly drained, the water falling from the clouds is at once absorbed, and passes downwards, saturating the soil in its descent, and carrying the soluble substances with it to the roots, and the surplus water runs away in the artificial channels provided by the draining process. So great is the absorbent power of drained land, that, after a protracted drought, all the water of a heavy rainstorm will be drunk up and held by the soil, so that, for a day or two, none will find its way to the drains, nor will it run upon the surface.

*Drainage lengthens the season for labor and vegetation.* In the colder latitudes of our country, where a long Winter is succeeded by a torrid Summer, with very little ceremony by way of an intervening

Spring, farmers have need of all their energy to get their seed seasonably into the ground. Snow often covers the fields in New England into April; and the ground is so saturated with water, that the land designed for corn and potatoes, frequently cannot be plowed till late in May. The manure is to be hauled from the cellar or yard, over land lifted and softened by frost, and all the processes of preparing and planting, are necessarily hurried and imperfect. In the Annual Report of the Secretary of the Board of Agriculture, of the State of Maine, for 1856, a good illustration of this idea is given: "Mr. B. F. Nourse, of Orrington, plowed and planted with corn a piece of his drained and subsoiled land, in a drizzling rain, after a storm of two days. The [262] corn came up and grew well; yet this was a clayey loam, formerly as wet as the adjoining grass-field, upon which oxen and carts could not pass, on the day of this planting, without cutting through the turf and miring deeply. The nearest neighbor said, if he had planted that day, it must have been from a raft." Probably two weeks would be gained in New England, in Spring, in which to prepare for planting, by thorough-drainage, a gain, which no one can appreciate but a New England man, who has been obliged often to plow his land when too wet, to cut it up and overwork his team, in hauling on his manure over soft ground, and finally to plant as late as the 6th of June, or leave his manure to waste, and lose the use of his field till another season; and all because of a surplus of cold water.

Mr. Yeomans, of New York, in a published statement of his experience in draining, says, that on his drained lands, "the ground becomes almost as dry in two or three days after the frost comes out in Spring, or after a heavy rain, as it would do in as many weeks, before draining." But the gain of time for labor is not all. We gain time also for vegetation, by thorough-drainage. Ten days, frequently, in New England, may be the security of our corn-crop against frost. In less than that time, a whole field passes from the milky stage, when a slight frost would ruin it, to the glazed stage, when it is safe from cold; and twice ten days of warm season are added by this removal of surplus water.

*Drainage prevents freezing out.* Mr. John Johnston, of Seneca County, New York, in 1851, had already made sixteen miles of tile drains. He had been experimenting with tiles from 1835, and had, on four acres of his drained clayey land, raised the largest crop of Indian corn ever produced in that county—eighty-three bushels of shelled corn to the acre. [263]

He states, that on this clayey soil, when laid down to grass, "not one square foot of the clover froze out." Again he says, "Heretofore, many acres of wheat were lost on the upland by freezing out, and none would grow on the lowlands. Now there is no loss from that cause."

The growing of Winter wheat has been entirely abandoned in some localities on account of freezing out, or Winter-killing; and one of the worst obstacles in the way of getting our lands into grass, and keeping them so, is this very difficulty of freezing out. The operation seems to be merely this: The soil is pulverized only to the depth of the plow, some six or eight inches. Below this is a stratum of clay, nearly impervious to water. The Autumn rains saturate the surface soil, which absorbs water like a sponge. The ground is suddenly frozen; the water contained in it crystallizes into ice; and the soil is thrown up into spicules, or honey-combs; and the poor clover roots, or wheat plants, are drawn from their beds, and, by a few repetitions of the process, left dead on the field in Spring. Draining, followed by subsoiling, lets down the falling water at once through the soil, leaving the root bed of the plants so free from moisture, that the earth is not "heaved," as the term is, and the plants retain their natural position, and awaken refreshed in the Spring by their Winter's repose.

*There are no open ditches on under-drained land.* An open ditch in a tillage or mowing-field, is an abomination. It compels us, in plowing, to stop, perhaps midway in our field; to make short lands; to leave headlands inconvenient to cultivate; and so to waste our time and strength in turning the team, and treading up the ground, instead of profitably employing it in drawing a long and handsome furrow the whole length of the field, as we might do were there no ditch. Open ditches, as usually made, obstruct the movement of our teams as [264] much as fences, and a farm cut into squares by ditches, is nearly as objectionable as a farm fenced off into half or quarter-acre fields.

In haying, we have the same inconvenience. We must turn the mowing-machine and horse-rake at the ditch, and finish by hand-labor, the work on its banks; we must construct bridges at frequent intervals, and then go out of our way to cross them with loads, cutting up the smooth fields with wheels and the feet of animals. Or, what is a familiar scene, when a shower is coming up, and the load is ready, Patrick concludes to drive straight to the barn, across the ditch, and gets his team mired, upsets his load, and perhaps breaks the leg of an animal, besides swearing more than half a mile of hard ditching will expiate. Such accidents are a great temptation to profanity, and under-draining might properly be reckoned a moral agent, to counteract such traps and pitfalls of the great adversary.

A moment's thought will satisfy any farmer who has the means, that true economy dictates a liberal expenditure of labor, at once, to obviate these difficulties, rather than be subject for a lifetime to the constant petty annoyances which have been named.

Open ditches, even when formed so skillfully that they may be conveniently crossed, or water-furrows which remain where land is laid into ridges by back-furrowing, as much of our flat land must be, if not under-drained, are serious obstructions, at the best.

They render the soil unequal in depth, taking it from one point where it is wanted, and heaping it upon another where it is not wanted, thus giving the crops an uneven growth. They render the soil also unequal in respect to moisture, because the back or top of the ridge must always be drier than the furrow.

Thorough-drained land may be laid perfectly flat, giving us, thus, the control of the whole field, to divide and cultivate [265] according to convenience, and making it of uniform texture and temperature.

Attempts have been made, to estimate the saving in the number of horses and men by drainage, and it is thought to be a reasonable calculation to fix it at one in four, or twenty-five per cent. It probably will strike any farmer as a fair estimate, that, on land which needs drainage, it will require four horses and four men to perform the same amount of cultivation, that three men and three horses may perform on the same land well drained.

*Drained land will not require re-planting.* There is hardly a farmer in New England, who does not, each Spring, find himself compelled to re-plant some portion of his crop. He is obliged to hurry his seed into the ground, at the earliest day, because our season for planting is short at the best. If, after this, a long cold storm comes, on wet land, the seed rots in the ground, and he must plant again, often too late, incurring thus the loss of the seed, the labor of twice doing the same work, the interruption of his regular plan of business, and often the partial failure of his crop.

Upon thorough-drained land, this cost and labor could rarely be experienced, because nothing short of a small deluge could saturate well drained land, so as to cause the seed to fail, if sowed or planted with ordinary care and prudence, as to the season.

*Drained land is lighter to work.* It is often difficult to find a day in the year, when a wet piece of land is in suitable condition to plow. Usually, such tracts are unequal, some parts being much wetter than others, because the water settles into the low places. In such fields, we now drive our team knee deep into soft mud, and find a stream of water following us in the furrow, and now we rise upon a knoll, baked hard, and sun-cracked; and one half the surface when finished is shining with the [266] plastered mud, ready to dry into the consistency of bricks, while the other is already in hard dry lumps, like paving stones, and about as easily pulverized.

This is hard work for the team and men, hard in the plowing, and hard through the whole rotation. The same field, well drained, is friable and porous, and uniform in texture. It may be well plowed and readily pulverized, if taken in hand at any reasonable season.

Land which has been puddled by the tread of cattle, or by wheels, acquires a peculiar consistency, and a singular capacity to hold water. Certain clays are wet and beaten up into this consistency, to form the bottoms of ponds, and to tighten dams and reservoirs. A soil thus puddled, requires careful treatment to again render it permeable to water, and fit for cultivation. This puddling process is constantly going on, under the feet of cattle, under the plow and the cart-wheels, wherever land containing clay is worked upon in a wet state. Thus, by performing a day's work on wet land, we often render necessary as much additional labor as we perform, to cure the evil we have done.

*We may haul loads without injury on drained land.* On many farms, it is difficult to select a season for hauling out manure, or carting stones from place to place, when great injury is not done to some part of the land by the operation. Many farmers haul out their manure in Winter, to avoid cutting up their farms; admitting that the manure is wasted somewhat by the exposure, but, on the whole, choosing this loss as the lesser evil. In spreading manure in Spring, we are often obliged to carry half loads, because the land is soft, not only to spare our beasts, but also to spare our land the injury by treading it. Drained land is comparatively solid, especially in Spring, and will bear up heavy loads with little injury. [267]

*Drained land is least injured by cattle in feeding.* Whether it is good husbandry to feed our mowing fields at any time, is a question upon which farmers have a right to differ. Without discussing the question, it is enough for our purpose, that most farmers feed their fields late in the Autumn. Whether we approve it, or not, when the pastures are bare and burnt up, and the second crop in the home-field is so rich and tempting, and the women are complaining that the cows give no milk, we usually bow to the necessity of the time, and "turn in" the cows. The great injury of "Fall-feeding" is not usually so much the loss of the grass-covering from the field, as the poaching of the soil and destruction of the roots by treading. A hard upland field is much less injured by feeding, than a low meadow, and the latter less in a dry than a wet season. By drainage, the surplus water is taken from the field. None can stand upon its surface for a day after the rain ceases. The soil is compact, and the hoofs of cattle make little impression upon it, and the second or third crop may be fed off, with comparatively little damage.

*Weeds are easily destroyed on drained land.* If a weed be dug or pulled up from land that is wet and sticky, it is likely to strike root and grow again, because earth adheres to its roots; whereas, a stroke of the hoe entirely separates the weeds in friable soil from the earth, and they die at once. Every farmer knows the different effect of hoeing, or of cultivating with the horse-hoe or harrow, in a rain storm and in dry weather. In one case, the weeds are rather refreshed by the stirring, and, in the other, they are destroyed. The difference between the surface of drained land and water-soaked land is much the same as that between land in dry weather under good cultivation, and land just saturated by rain.

Again, there are many noxious weeds, such as wild grasses, which thrive only on wet land, and which are [268] difficult to exterminate, and which give us no trouble after the land is lightened and sweetened by drainage. Among the effects of drainage, mainly of a chemical nature, on the soil, are the following:

*Drainage promotes absorption of fertilizing substances from the air.* The atmosphere bears upon its bosom, not only the oxygen essential to the vitality of plants, not only water in the form of vapor, to quench their thirst in Summer droughts, but also various substances, which rise in exhalations from the sea, from decomposing animals and vegetables, from the breathing of all living creatures, from combustion, and a thousand other causes. These would be sufficient to corrupt the very air, and render it unfit for respiration, did not Nature, with her wondrous laws of compensation, provide for its purification. It has already been stated, how the atmosphere returns to the hills, in clouds and vapor, condensed at last to rain, all the water which the rivers carry to the sea; and how the well-drained soil derives moisture, in severest time of need, from its contact with the vapor-loaded air. But the rain and dew return not their waters to the earth without treasures of fertility. Ammonia, which is one of the most valuable substances found in farm-yard manures, and which is a constant result of decomposition, is absorbed in almost incredible quantities by water. About 780 times its own bulk of ammonia is readily absorbed by water at the common temperature and pressure of the atmosphere; and, freighted thus with treasures for the fields, the moisture of the atmosphere descends upon the earth. The rain cleanses the air of its impurities, and conveys them to the plants. The vapors of the marshes, and of the exposed manure heaps of the thriftless farmer, are gently wafted to the well-drained fields of his neighbor, and there, amidst the roots of the well-tilled crops, deposit, at the same time, their moisture and fertilizing wealth. [269]

Of the wonderful power of the soil to absorb moisture, both from the heavens above and the earth beneath—by the deposition of dew, as well as by attraction—we shall treat more fully in another chapter. It will be found to be intimately connected with the present topic.

*Thorough drainage supplies air to the roots.* Plants, if they do not breathe like animals, require for their life almost the same constant supply of air. "All plants," says Liebig, "die in soils and water destitute of oxygen; absence of air acts exactly in the same manner as an excess of carbonic acid. Stagnant water on a marshy soil excludes air, but a renewal of water has the same effect as a renewal of air, because water contains it in solution. When the water is withdrawn from a marsh, free access is given to the air, and the marsh is changed into a fruitful meadow." Animal and vegetable matter do not decay, or decompose, so as to furnish food for plants, unless freely supplied with oxygen, which they must obtain from air. A slight quantity of air, however, is sufficient for putrefaction, which is a powerful deoxydizing process that extracts oxygen even from the roots of plants.

We are accustomed to think of the earth as a compact body of matter, vast and inert; subject, indeed, to be upheaved and rent by volcanoes and earthquakes, but as quite insensible to slight influences which operate upon living beings and upon vegetation. This, however, is a great mistake; and it may be interesting to refer to one or two facts, which illustrate the wonderful effect of changes of the atmosphere upon the soil, and upon the subterranean currents of the earth. The following is from remarks by Mr. Denton, in a public address:

"But, as a proof of the sensibility of a soil drained four feet deep, to atmospheric changes, I may mention, that my attention has been, on more than one occasion, called to the circumstance that drains have been observed to run, after a discontinuance of that duty, without any [270] fall of rain on the surface of the drained land; and, upon reference to the barometer, it has been found that the quicksilver has fallen whenever this has occurred. Mr. George Beaumont, jun., who first afforded tangible evidence of this extraordinary circumstance, has permitted me to read the following extracts of his letter:

"'I can verify the case of the drains running without rain, during a falling barometer, beyond all doubt.

"'The case I named to you last year of the barometer falling four days consecutively, and with rapidity, was a peculiarly favorable time for noticing it, as it occurred in a dry time, and the drains could be seen distinctly. My man, on being questioned and cautioned by me not to exaggerate, has declared the actual stream of water issuing from one particular drain to be as thick as a three-eighth-inch wire. All the drains ran—they did more than drop—and ditches, which were previously dry, became quite wet, with a perceptible stream of water; this gradually ceased with the change in the density of the atmosphere, as shown by the barometer.

"'During last harvest, 1855, the men were cutting wheat, and on getting near to a drain outlet, the ditch from the outlet downwards was observed to be wet, and the drain was dripping. No rain fell in sufficient quantity to enter the ground. The men drank of the water while they were cutting the wheat. A few days after, it was dry again. I have seen and noticed this phenomenon myself.'

"A correspondent of the *Agricultural Gazette* has stated, that Professor Brocklesby, of Hartford, in America, had observed the same phenomena, in the case of two springs in that country; and explained, that the cause was 'the diminished atmospheric pressure which exists before a rain.'"

Dr. Lardner states many facts which support the ideas above suggested. In his lectures on science, he says: "When storms are breaking in the heavens, and sometimes long before their commencement, and when their approach has not yet been manifested by any appearances in the firmament, phenomena are observed, apparently sympathetic, proceeding from the deep recesses of the earth, and exhibited under very various forms at its surface." Dr. Lardner cites many instances of fountains which, when a storm is approaching, burst forth with a violent flow of water, before any rain has fallen. [271]

The cases named by Prof. Brocklesby, referred to by Mr. Denton, are those of a spring in Rutland, Vermont, and a brook in Concord, Massachusetts. Prof. Brocklesby states, as the result of his personal observation, that the spring referred to, supplies an aqueduct; that, in several instances, when the spring had become so low, in a time of drought, that no water ran in the aqueduct, it suddenly rose so as to fill the pipes, and furnish a supply of water, before any rain had fallen in the neighborhood. This occurrence, he says, was familiar to the occupants of the premises, and they expected rain in a few days after this mysterious flow of water; which expectations were usually, if not always, realized.

The other instance is that of a brook in Concord, Mass., called Dodge's brook, which Prof. B. says, he was informed, commenced frequently to rise very perceptibly before a drop of rain had fallen.

We have inquired of our friends in Concord about this matter, and find that this opinion is entertained by many of the people who live near this brook, and it is probably well founded, though we cannot ascertain that accurate observations have been made, so as to afford any definite results.

*Thorough drainage warms the soil.* It has been stated, on high authority, that drainage raises the temperature of the soil, often as much as 15° F. Indian corn vegetates at about 55°. At 45°, the seed would rot in the ground, without vegetating. The writer, however, has seen rye sprouted upon ice in an ice-house, with roots two inches long, so grown to the ice that they could only be separated by thawing. Winter rye, no doubt, makes considerable growth under snow. Cultivated plants, in general, however, do not grow at all, unless the soil be raised above 45°. The sun has great power to warm dry soils, and, it is said, will often raise their temperature to 90° or 100°, [272] when the air in the shade is only 60° or 70°. But the sun has no such power to warm a wet soil, and for several reasons, which are as follows:

1. *The soil is rendered cold by evaporation.* If water cannot pass through the land by drainage, either natural or artificial, it must escape, if at all, at the surface, by evaporation. Now, it is a fact well known, that the heat disappears, or becomes latent, by the conversion of water into vapor. Every child knows this, practically, at least,

who, in Winter, has washed his hands and gone out without drying them. The same evaporation which thus affects the hands, renders the land cold, when filled with water, every gallon of which thus carried off requires, and actually carries off, as much heat as would raise five and a half gallons of water from the freezing to the boiling point.

Morton, in his "Encyclopædia of Agriculture," estimates that it would require an expenditure of nearly 1,200 pounds of coal per day, to evaporate artificially one half the rain which falls on an acre during the year. In other words, about 219 tons of coals annually, would be required for every acre of undrained land, so as to allow the free use of the sun's rays for the legitimate purpose of growing and maturing the crops cultivated upon it. It will not then be surprising that undrained soils are, in the language of the farmer, "cold."

2. *Heat will not pass downward in water.* If, therefore, your soil be saturated with water, the heat of the sun, in Spring, cannot warm it, and your plowing and planting must be late, and your crop a failure. Count Rumford tried many experiments to illustrate the mode of the propagation of heat in fluids, and his conclusion, it is presumed, is now held to be the true theory, that heat is transmitted in water only by the motion of the particles of water; so that, if you could stop the heated particles [273] from rising, water could not be warmed except where it touches the vessel containing it. Heat applied to the bottom of a vessel of water warms the particles in contact with the vessel, and colder particles descend, and so the whole is warmed.

Heat, applied to the surface of the water, can never warm it, except so far as it is conducted downward by some other medium than the water itself. Count Rumford confined cakes of ice in the bottom of glass jars, and, covering it with one thickness of paper, poured boiling-hot water on the top of it, and there it remained for hours without melting the ice. The paper was placed over the ice, so that the hot water could not be poured on it, which would have thawed it at once. Every man who has poured hot water into a frozen pump, hoping to thaw out the ice by this means, has arrived at the fact, if not at the theory, that ice will not melt by hot water on the top of it. If, however, a piece of lead pipe be placed in the pump, resting on the ice, and hot water be poured through it, the ice will melt at once. In the first instance, the hot water in contact with the ice becomes cold; and there it remains, because cold water is heavier than warm, and there it will remain, though the top be boiling. But when hot water is poured through the pipe, the downward current drives away the cold water, and brings heated particles in succession to the ice.

Heat is propagated in water, then, only by circulation; that is, by the upward movement of the heated particles, and the downward movement of the colder ones to take their place. Anything which obstructs circulation, prevents the passage of heat. Chocolate retains heat longer than tea, because it is thicker, and the hot particles cannot so readily rise to be cooled at the surface. Count Rumford illustrated this fact satisfactorily, by putting eider-down into water, which was found to obstruct the [274] circulation, and to prevent the rapid heating and cooling of it. The same is true of all viscous substances, as starch and glue; and so of oil. They retain heat much longer than water or spirits.

In a soil saturated with water, or even in water thickened with mud, there could then be but little circulation of the particles, even were the heat applied at the bottom instead of the top. Probably the soil, though saturated with water, does, to some extent, transmit heat from one particle of earth to another, but it must be but very slowly.

In the chapter upon Temperature as affected by Drainage, farther illustrations of this point may be found.

## AERATION BY DRAINS.

Among the advantages of thorough-drainage, is reckoned by all, the circulation of air through the soil. No drop of water can run from the soil into a drain without its place being supplied by air, unless there is more water to supply it; so that drainage, in this way, manifestly promotes the permeation of air through the soil.

But it is claimed that drains may be made to promote circulation of air in another way, and in dry times, when no water is flowing through them, by connecting them together by means of a header at the upper ends, and leaving an opening so that the air may pass freely through the whole system. Our friend, Prof. Mapes, is an advocate for this practice, and certainly the theory seems well supported. It is said that in dry, hot weather, when the air is most highly charged with moisture, currents thus passing constantly through the earth, must, by contact with the cooler subsoil, part with large quantities of moisture, and tend to moisten the soil from the drains to the surface, giving off also with the moisture whatever of fertilizing elements the air may bear with it.

This point has not escaped the notice of English drainers. [275] Mr. J. H. Charnock, an assistant commissioner under the Drainage act, in 1843, read a paper in favor of this practice, but in 1849 he published a second article in which he suggests doubts of the advantages of such arrangements, and says he has discontinued their application. He says they add to the cost of the work, and tend to the decay of the pipes, and to promote the growth into the pipes, of any roots that may approach them.

Mr. Parkes, in a published article in 1846, speaks of this idea, but passes it by as of very little importance. Mr. Denton quotes the authority of some of his correspondents strongly in favor of this theory. After trying some experiments himself upon clay soil, he admits the advantages of such an arrangement for such soil, in the following not very enthusiastic terms:

"It will be readily understood that as clay will always contract rapidly under the influence of a draught of air, in consequence of the rapid evaporation of moisture from its surface, one of the benefits of draining is thus very cheaply acquired; and for the denser clays it may possibly be a desirable thing to do, but in the porous soils it would appear that no advantage is gained by it."

Yet, notwithstanding this summary disposition of the question in England, it is by no means clear, that in the tropical heat of American summers, when the difference between the temperature of the air and the subsoil is so much greater than it can ever be in England, and when we suffer from severer droughts than are common there, we may not find substantial practical advantage from the passage of these air currents through the soil.

We are not aware of experiments in America, accurate enough to be quoted as authority on the subject.

# CHAPTER XIV [276]
# DRAINAGE ADAPTS THE SOIL TO GERMINA-
# TION AND VEGETATION.

Process of Germination.—Two Classes of Pores in Soils, illustrated by Cuts.—Too much Water excludes Air, reduces Temperature.—How much Air the Soil Contains.—Drainage Improves the Quality of Crops.—Drainage prevents Drought.—Drained Soils hold most Water.—Allow Roots to go Deep.—Various Facts.

No apology will be necessary for the long extract which we are about to give, to any person who will read it with attention. It is from a lecture on Agricultural Science, by Dr. Madden, and we confess ourselves incompetent to condense or improve the language of the learned author.

We think we are safe in saying that it has never been before published in America:

"The first thing which occurs after the sowing of the seed is, of course, *germination*; and before we examine how this process may be influenced by the condition of the soil, we must necessarily obtain some correct idea of the process itself. The most careful examination has proved that the process of germination consists essentially of various chemical changes, which require for their development the presence of air, moisture, and a certain degree of warmth. Now it is obviously unnecessary for our present purpose that we should have the least idea of the nature of these processes: all we require to do, is to ascertain the conditions under which they take place; having detected these, we know at once what is required to make a seed grow. These, we have seen, are air, moisture, and a certain degree of warmth; and it consequently results, that wherever a seed is placed in these circumstances, germination will take place. Viewing matters in this light, it appears [277] that soil does not act *chemically* in the process of germination; that its sole action is confined to its being the vehicle, by means of which a supply of air and moisture and warmth can be continually kept up. With this simple statement in view, we are quite prepared to consider the various conditions of soil, for the purpose of determining how far these will influence the future prospects of the crop, and we shall accordingly at once proceed to examine carefully into the *mechanical relations of the soil.* This we propose doing by the aid of figures. Soil examined mechanically, is found to consist entirely of particles of all shapes and sizes, from stones and pebbles, down to the finest powder; and, on account of their extreme irregularity of shape, they cannot lie so close to one another as to prevent there being passages between them, owing to which circumstance soil in the mass is always more or less *porous*. If, however, we proceed to examine one of the smallest particles of which soil is made up, we shall find that even this is not always solid, but is much more frequently porous, like soil in the mass. A considerable proportion of this finely-divided part of soil, *the impalpable matter* as it is generally called, is found, by the aid of the microscope, to consist of *broken-down vegetable tissue*, so that when a small portion of the finest dust from a garden or field is placed under the microscope, we have exhibited to us particles of every variety of shape and structure, of which a certain part is evidently of vegetable origin. In these figures I have given a very rude representation of these particles; and I must beg you particularly to remember that they are not meant to represent by any means accurately what the microscope exhibits, but are only designed to serve as a plan by which to illustrate the mechanical properties of the soil. On referring to Fig. 91, we perceive that there are two distinct classes of pores; first, the large ones, which exist *between* the particles of soil, and second, the very minute ones, which occur in the particles themselves; and you will at the same time notice, that whereas all the larger pores—those between the [278] particles of soil—communicate most freely with each other, so that they form canals, the small pores, however freely they may communicate with one another in the interior of the particle in which they occur, have no direct connection with the pores of the surrounding particles. Let us now, therefore, trace the effect of this arrangement. In Fig. 91, we perceive that these canals and pores are all empty, the soil being *perfectly dry*; and the canals communicating freely at the surface with the surrounding atmosphere, the whole will of course be filled with air. If in this condition, a seed be placed in the soil, as at *a*, you at once perceive that it is freely supplied with air, *but there is no moisture*; therefore, when soil is *perfectly dry*, a seed cannot grow.

Fig. 91.

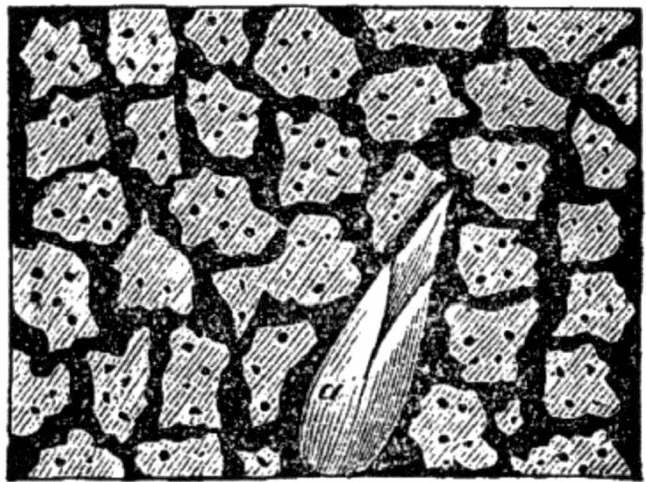

Fig. 92.

"Let us turn our attention now to Fig. 92. Here we perceive that both the pores and canals are no longer represented white, but black, this color being used to indicate water; in this instance, therefore, water has taken the place of air, or, in other words, the soil is *very wet*. If we observe our seed *a* now, we find it abundantly supplied with water, but *no air*. Here again, therefore, germination cannot take place. It may be well to state here, that this can never occur *exactly* in nature, because water having the power of dissolving air to a certain extent, the seed *a* in Fig. 92 is, in fact, supplied with a *certain* amount of this necessary substance; and, owing to this, germination does take place, although by no means under such advantageous circumstances as it would were the soil in a better condition.

Fig. 93.

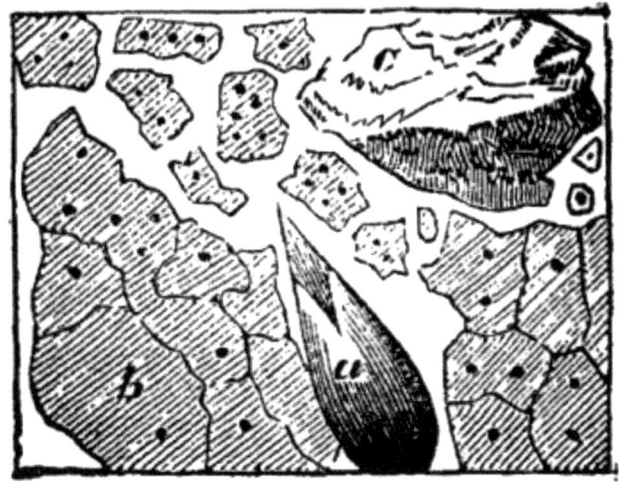

Fig. 94.

"We pass on now to Fig. 93. Here we find a different state of matters. The canals are open and freely supplied with air, while the pores are filled with water; and consequently you perceive that, while the seed *a* has quite enough of air from the canals, it can never be without moisture, as every particle of soil which touches it, is well supplied with this necessary ingredient. This, then, is the proper condition of soil for germination, and in fact for every period of the plant's development; and this condition occurs when soil is *moist* but not *wet* — that [279] is to say, when it has the color and appearance of being well watered, but when it is still capable of being crumbled to pieces by the hands, without any of its particles adhering together in the familiar form of mud.

"Turning our eyes to Fig. 94, we observe still another condition of soil. In this instance, as far as *water* is concerned, the soil is in its healthy condition — it is moist, but not wet, the pores alone being filled with water. But where are the canals? We see them in a few places, but in by far the greater part of the soil none are to be perceived; this is owing to the particles of soil having adhered together, and thus so far obliterated the interstitial canals, that they appear only like pores. This is the state of matters in every *clod of earth, b*; and you will at once perceive, on comparing it with *c*, which represents a stone, that these two differ only in possessing a few pores, which latter, while they may form a reservoir for moisture, can never act as vehicles for the *food* of plants, as the roots are not capable of extending their fibres into the interior of a clod, but are at all times confined to the interstitial canals.

"With these four conditions before us, let us endeavor to apply them *practically* to ascertain when they occur in our fields, and how those which are injurious may be obviated.

"The first of them, we perceive, is a state of too great dryness, *a very rare* condition, in this climate at least; in fact, the only case in which it is likely to occur is in very coarse sands, where the soil, being chiefly made up of pure sand and particles of flinty matter, contains comparatively much fewer pores; and, from the large size of the individual particles, assisted by their irregularity, the canals are wider, the circulation of air freer, and, consequently, the whole is much more easily dried. When this state of matters exists, the best treatment is to leave all the stones which occur on the surface of the field, as they cast shades, and thereby prevent or retard the evaporation of water.

"We will not, however, make any further observations on this very rare case, but will rather proceed to Fig. 92, a much more frequent, and, in every respect, more important condition of soil: I refer to an *excess of water*.

"When water is added to perfectly dry soil, it, of course, in the first instance, fills the interstitial canals, and from these enters the pores of each particle; and if the supply of water be not too great, the canals speedily become empty, so that the whole of the fluid is taken up by the pores: this, we have already seen, is the *healthy* condition of the soil. If, however, the supply of water be too great, as is the case when [280] a spring gains admission into the soil, or when the sinking of the fluid through the canals to a sufficient depth below the surface is prevented, it is clear that these also must get filled with water so soon as the pores have become saturated. This, then, is the condition of *undrained soil*.

"Not only are the pores filled, but the interstitial canals are likewise full; and the consequence is, that the whole process of the germination and growth of vegetables is materially interfered with. We shall here, therefore briefly state the injurious effects of an excess of water, for the purpose of impressing more strongly on your minds the necessity of thorough-draining, as the first and most essential step towards the improvement of your soil.

"The *first* great effect of an excess of water is, that it produces a corresponding diminution of the amount of air beneath the surface, which air is of the greatest possible consequence in the nutrition of plants; in fact, if entirely excluded, germination could not take place, and the seed sown would, of course, either decay or lie dormant.

"*Secondly*, an excess of water is most hurtful, by reducing considerably the *temperature* of the soil: this I find, by careful experiment, to be to the extent of six and a-half degrees Fahrenheit in Summer, which amount is equivalent to an elevation above the level of the sea of 1,950 feet.

"These are the two chief injuries of an excess of water in soil which affect the soil itself. There are very many others affecting the climate, &c.; but these not so connected with the subject in hand as to call for an explanation here.

"Of course, all these injurious effects are at once overcome by thorough-draining, the result of which is, to establish a direct communication between the interstitial canals and the drains, by which means it follows, that no water can remain any length of time in these canals without, by its gravitation, finding its way into the drains.

"The 4th Fig. indicates badly cultivated soil, or soil in which large unbroken clods exist; which clods, as we have already seen, are very little better than stones, on account of their impermeability to air and the roots of plants.

"Too much cannot be said in favor of pulverizing the soil; even thorough-draining itself will not supersede the necessity of performing this most necessary operation. The whole valuable effects of plowing, harrowing, grubbing, &c., may be reduced to this: and almost the whole superiority of *garden* over *field* produce is referable to the greater perfection to which this pulverizing of the soil can be carried. [281]

"The whole success of the drill husbandry is owing, in a great measure, to its enabling you to stir up the soil well during the progress of your crop; which stirring up is of no value beyond its effects in more minutely pulverizing the soil, increasing, as far as possible, the size and number of the interstitial canals.

"Lest any one should suppose that the contents of these interstitial canals must be so minute that their whole amount can be of but little consequence, I may here notice the fact, that, in moderately well pulverized soil, they amount to no less than one-fourth of the whole bulk of the soil itself; for example, 100 cubic inches of *moist* soil (that is, of soil in which the pores are filled with water while the canals are filled with air), contain no less than 25 cubic inches of air. According to this calculation, in a field pulverized to the depth of eight inches, a depth perfectly attainable on most soils by careful tillage, every imperial acre will retain beneath its surface no less than 12,545,280 cubic inches of air. And, to take one more element into the calculation, supposing the soil were not properly drained, the sufficient pulverizing of an additional inch in depth would increase the escape of water from the surface by upwards of one hundred gallons a day."

*Drainage improves the quality of crops.* In a dry season, we frequently hear the farmer boast of the quality of his products. His hay-crop, he says, is light, but will "spend" much better than the crop of a wet season; his potatoes are not large, but they are sound and mealy. Indeed, this topic need not be enlarged upon. Every farmer knows that his wheat and corn are heavier and more sound when grown upon land sufficiently drained.

*Drainage prevents drought.* This proposition is somewhat startling at first view. How can draining land make it more moist? One would as soon think of watering land to make it dry. A drought is the enemy we all dread. Professor Espy has a plan for producing rain, by lighting extensive artificial fires. A great objection to his theory is, that he cannot limit his showers to his own land, and all the public would never be ready for a shower on the same day. If we can really protect our land from drought, by under-draining it, everybody may at once engage in the work without offence to his neighbor. [282]

If we take up a handfull of rich soil of almost any kind, after a heavy rain, we can squeeze it hard enough with the hand to press out drops of water. If we should take of the same soil a large quantity, after it was so dry that not a drop of water could be pressed out by hand, and subject it to the pressure of machinery, we should force from it more water. Any boy, who has watched the process of making cider with the old-fashioned press, has seen the pomace, after it had been once pressed apparently dry and cut down, and the screw applied anew to the "cheese," give out quantities of juice. These facts illustrate, first, how much water may be held in the soil by attraction. They show, again, that more water is held by a pulverized and open soil, than by a compact and close one. Water is held in the soil between the minute particles of earth. If these particles be pressed together compactly, there is no space left between them for water. The same is true of soil naturally compact. This compactness exists more or less in most subsoils, certainly in all through which water does not readily pass. Hence, all these subsoils are rendered more permeable to water by being broken up and divided; and more retentive by having the particles of which they are composed separated, one from another—in a word, by pulverization. This increased capacity to contain moisture by attraction, is the greatest security against drought. The plants, in a dry time send their rootlets throughout the soil, and flourish in the moisture thus stored up for their time of need. The pulverization of drained land may be produced, partly by deep, or subsoil plowing, which is always necessary to perfect the object of thorough-draining; but it is much aided, in stiff clays, also, by the shrinkage of the soil by drying.

Drainage resists drought, again, by the very deepening of the soil of which we have already spoken. The roots [283] of plants, we have seen, will not extend into stagnant water. If, then, as is frequently the case, even on sandy plains, the water-line be, in early Spring, very near the surface, the seed may be planted, may vegetate, and throw up a goodly show of leaves and stalks, which may flourish as long as the early rains continue; but, suddenly, the rains cease; the sun comes out in his June brightness; the water-line lowers at once in the soil; the roots have no depth to draw moisture from below, and the whole field of clover, or of corn, in a single week, is past recovery. Now, if this light, sandy soil be drained, so that, at the first start of the crop, there is a deep seed-bed free from water, the roots strike downward, at once, and thus prepare for a drought. The writer has seen upon deep-trenched land in his own garden, parsnips, which, before midsummer, had extended downward three feet, before they were as large as a common whiplash; and yet, through the Summer drought, continued to thrive till they attained in Autumn a length, including tops, of about seven feet, and an extraordinary size. A moment's reflection will satisfy any one that, the dryer the soil in Spring, the deeper will the roots strike, and the better able will be the plant to endure the Summer's drought.

Again, drainage and consequent pulverization and deepening of the soils increase their capacity to absorb moisture from the atmosphere, and thus afford protection against drought. Watery vapor is constantly, in all dry weather, rising from the surface of the earth; and plants, in the day-time, are also, from their leaves and bark, giving off moisture which they draw from the soil. But Nature has provided a wonderful law of compensation for this waste, which would, without such provision, parch the earth to barrenness in a single rainless month.

The capacity of the atmosphere to take up and convey water, furnishes one of the grandest illustrations of the [284] perfect work of the Author of the Universe. "All the rivers run into the sea, yet the sea is not full;" and the sea is not full, because the numerous great rivers and their millions of tributaries, ever flowing from age to age, convey to the ocean only as much water as the atmosphere carries back in vapor, and discharges upon the hills. The warmer the atmosphere, the greater its capacity to hold moisture. The heated, thirsty air of the tropics drinks up the water of the ocean, and bears it away to the colder regions, where, through condensation by cold, it becomes visible as a cloud; and as a huge sponge pressed by an invisible hand, the cloud, condensed still further by cold, sends down its water to the earth in rain.

The heated air over our fields and streams, in Summer, is loaded with moisture as the sun declines. The earth has been cooled by radiation of its heat, and by constant evaporation through the day. By contact with the cooler soil, the air, borne by its thousand currents gently along its surface, is condensed, and yields its moisture to the thirsty earth again, in the form of dew.

At a Legislative Agricultural Meeting, held in Albany, New York, January 25th, 1855, "the great drought of 1854" being the subject, the secretary stated that "the experience of the past season has abundantly proved that thorough-drainage upon soils requiring it, has proved a very great relief to the farmer;" that "the crops upon such lands have been far better, generally, than those upon undrained lands, in the same locality;" and that, "in many instances, the increased crop has been sufficient to defray the expenses of the improvement in a single year."

Mr. Joseph Harris, at the same meeting, said: "An underdrained soil will be found damper in dry weather, than an undrained one, and the thermometer shows a drained [285] soil warmer in cold weather, and cooler in hot weather, than one which is undrained."

The secretary of the New York State Agricultural Society, in his Report for 1855, says: "The testimony of farmers, in different sections of the State, is almost unanimous, that drained lands have suffered far less from drought than undrained." Alleghany county reports that "drained lands have been less affected by the drought than undrained;" Chatauque county, that "the drained lands have stood the drought better than the undrained." The report from Clinton county says: "Drained lands have been less affected by the drought than undrained." Montgomery county reports: "We find that drained lands have a better crop in either wet or dry seasons than undrained."

B. F. Nourse, of Orrington, Maine, states that, on his drained land, in that State, "during the drought of 1854, there was at all times sufficient dampness apparent on scraping the surface of the ground with his foot in passing, and a crop of beans was planted, grown and gathered therefrom, without as much rain as will usually fall in a shower of fifteen minutes' duration, while vegetation on the next field was parching for lack of moisture."

A committee of the New York Farmers' Club, which visited the farm of Prof. Mapes, in New Jersey, in the time of a severe drought, in 1855, reported that the Professor's fences were the boundaries of the drought, all the lands outside being affected by it, while his remained free from injury. This was attributed, both by the committee and by Prof. Mapes himself, to thorough-drainage and deep tillage with the subsoil plow.

Mr. Shedd, in the *N. E. Farmer*, says:

"A simple illustration will show the effect which stagnant water, within a foot or two of the surface, has on the roots of plants. [286]

"Perhaps it will aid the reader, who doubts the benefit of thorough-draining in case of drought, to see why it is beneficial.

Fig. 95.

Fig. 96.

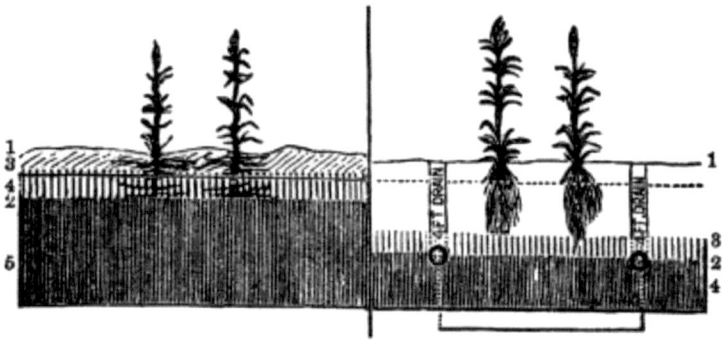

Section of land before it is drained.

Section of land after it is drained.

"In the first figure, 1 represents the surface soil, through which evaporation takes place, using up the heat which might otherwise go to the roots of plants; 2, represents the water table, or surface of stagnant water below which roots seldom go; 3, water of evaporation; 4, water of capillary attraction; 5, water of drainage, or stagnant water.

"In the second figure, 1 represents the surface-soil warmed by the sun and Summer rains; 2, the water-table nearly four feet below the surface—roots of the wheat plant have been traced to a depth of more than four feet in a free mold; 3, water of capillary attraction; 4, water of drainage, or stagnant water."

# CHAPTER XV [287]
# TEMPERATURE AS AFFECTED BY DRAINAGE.

Drainage Warms the Soil in Spring.—Heat cannot go down in Wet Land.—Drainage causes greater Deposit of Dew in Summer.—Dew warms Plants in Night, Cools them in the Morning Sun.—Drainage varies Temperature by Lessening Evaporation.—What is Evaporation.—How it produces Cold.—Drained Land Freezes Deepest, but Thaws Soonest, and the Reasons.

*Drainage raises the temperature of the soil, by allowing the rain to pass downwards.* In the growing season, especially in the Spring, the rain is considerably warmer than the soil. If the soil be saturated with the cold snow-water, the water which falls must, of course, run away upon the surface. If the soil be drained, the rain-water finds ready admission into it, carrying and imparting to it a portion of its heat. The experiments of Count Rumford, showing that heat is not propagated downward in fluids, may be found at page 273. This is a principle too important to be overlooked, especially in New England, where we need every aid from Nature and Art, to contend successfully against the brevity of the planting season. Soil saturated with cold water, cannot be warmed by any amount of heat applied to the surface. Warm water is lighter than cold water, and stays at the surface. In boiling water in a kettle, we apply fire at the bottom, and no amount of heat at the surface of the vessel would produce the desired effect. So rapid is the passage of heat upward in water, that the hand may without injury be held upon the bottom of a kettle of boiling water one minute after it has been removed from the fire. [288]

The following experiments and illustrations, from the *Horticulturist* of Nov. 1856, beautifully illustrate this point:

## "RATIONALE OF DRAINING LAND EXPLAINED.

"The reason why drained land gains heat, and water-logged land is always cold, consists in the well-known fact that heat cannot be transmitted *downwards* through water. This may readily be seen by the following experiments:

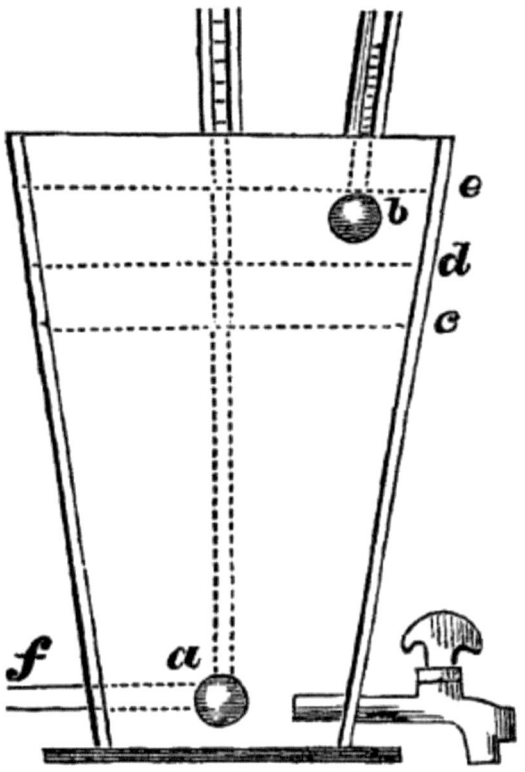

Fig. 97.

"*Experiment No. 1.*—A square box was made, of the form represented by the annexed diagram, eighteen inches deep, eleven inches
wide at top, and six inches wide at bottom. It was filled with peat,
saturated with water to *c*, forming to that depth (twelve and a half
inches) a sort of artificial bog. The box was then filled with water to
*d*. A thermometer *a*, was plunged, so that its bulb was within one
inch and a half of the bottom. The temperature of the whole mass of
peat and water was found to be 39½° Fahr. A gallon of boiling water was then added; it raised the surface of the water to *e*. In five
minutes, the thermometer, *a*, rose to 44°, owing to the conduction of
heat by the thermometer and its guard tube; at ten minutes from the

introduction of the hot water, the thermometer, *a*, rose to 46°, and it subsequently rose no higher. Another thermometer, *b*, dipping under the surface of the water at *e*, was then introduced, and the following are the indications of the two thermometers at the respective intervals, reckoning from the time the hot water was supplied:

|  |  |  |  | Thermometer b. | Thermometer a. |
|---|---|---|---|---|---|
|  | 20 | minutes |  | 150° | 46° |
| 1 hour | 30 | " |  | 101° | 45° |
| 2 hours | 30 | " |  | 80½° | 42° |
| 12 " | 40 | " |  | 45° | 40° |

"The mean temperature of the external air to which the box was exposed [289] during the above period, was 42°, the maximum being 47°, and the minimum 37°.

"*Experiment No. 2.* — With the same arrangement as in the preceding case, a gallon of boiling water was introduced above the peat and water, when the thermometer *a*, was at 36°; in ten minutes it rose to 40°. The cock was then turned for the purpose of drainage, which was but slowly effected; and, at the end of twenty minutes, the thermometer *a*, indicated 40°; at twenty-five minutes, 42°, whilst the thermometer *b*, was 142°. At thirty minutes, the cock was withdrawn from the box, and more free egress of water being thus afforded, at thirty-five minutes the flow was no longer continuous, and the thermometer *b*, indicated 48°. The mass was drained, and permeable to a fresh supply of water. Accordingly, another gallon of boiling water was poured over it; and, in

| 3 minutes, the thermometer *a*, rose to | 77°. |
|---|---|
| 5 minutes, the thermometer *a*, fell to | 76½°. |
| 15 minutes, the thermometer *a*, fell to | 70½°. |
| 20 minutes, the thermometer *a*, remained at | 71°. |

1 hour 50 minutes, the thermometer *a*, remained at 70½°.

"In these two experiments, the thermometer at the bottom of the box suddenly rose a few degrees immediately after the hot water was added; and it might be inferred that the heat was carried downwards by the water. But, in reality, the rise was owing to the action of the hot water on the thermometer, and not to its action upon the cold water. To prove this, the perpendicular thermometers were removed. The box was filled with peat and water to within three inches of the top, a horizontal thermometer, *a f*, having been previously secured through a hole made in the side of the box, by means of a tight-fitting cork, in which the naked stem of the thermometer was grooved. A gallon of boiling water was then added. The thermometer, a very delicate one, was *not in the least affected* by the boiling water in the top of the box.

"In this experiment, the wooden box may be supposed to be a field; the peat and cold water represent the water-logged portion; rain falls on the surface, and becomes warmed by contact with the soil, and, thus heated, descends. But it is stopped by the cold water, and the heat will go no further. But, if the soil is drained, and not water-logged, the warm rain trickles through the crevices of the earth, carrying to the drain-level the high temperature it had gained on the surface, parts [290] with it to the soil as it passes down, and thus produces that bottom heat which is so essential to plants, although so few suspect its existence."

Water, although it will not conduct heat downwards, is a ready vehicle of cold from the surface towards the bottom. Water becomes heavier by cooling till it is reduced to about 39°, at which point it attains its greatest density, and has a tendency to go to the bottom until the whole mass is reduced to this low temperature. Thus, the circulation of water in the saturated soil, in some conditions of the temperature of the surface and subsoil, may have a chilling effect which could not be produced on drained soil.

After water is reduced to about 39°, instead of obeying the common law of becoming heavier by cooling, it forms a remarkable exception to it, and becomes lighter until it freezes. Were it not for this admirable provision of Nature, all our ponds and rivers would, in the Winter, become solid ice from the surface to the bottom. Now as the surface water is chilled it goes to the bottom, and is replaced by warmer water, which rises, until the whole is reduced to the point of greatest density. Then the circulation ceases, and the water colder than 39° remains at the surface, is converted into ice which becomes still lighter, by crystallization, and floats upon the surface.

No experiments, showing the temperature of undrained soils at various depths, in the United States, have come to our knowledge. Mr. Gisborne says: "Many experiments have shown that, in retentive soils, the temperature, at two or three feet below the surface of the water-table, is, at no period of the year, higher than from 46° to 48° in agricultural Britain." Prof. Henry states in the Patent Office Report for 1857, that in the cellars of the observatory, at Paris, at the depth of sixty-seven and a half feet, in fifty years, the temperature has never varied a tenth of a degree from 53° 28', in all that period, Summer or Winter. [291]

Mr. Parkes gives the results of a valuable series of experiments, in which he compared the temperature of drained and undrained portions of a bog. He found the temperature of the undrained portion to remain steadily at 46°, at all depths, from one to thirty feet; and at seven inches from the surface, the temperature remained at 47° during the experiments. During the same period, the temperature of the drained portion was 48¼° at two feet seven inches below the surface, and at seven inches, reached as high as 66° during a thunder-storm; while, on a mean of thirty-five observations, the temperature at the latter depth was 10° higher than at the same depth in the undrained portion of the bog.

We find in the "Agriculture of New York," the results of observations made at Albany and at Scott, in that State, in the year 1848, upon temperature at different depths. The condition of the soil is not described, but it is presumed that it was soil naturally drained in both cases. A few of the results may give the reader some idea of the range of underground temperature, as compared with that of the air.

Temperature at Albany at two feet depth.

|   |   |   |   |   |
|---|---|---|---|---|
| " | " | " | highest August 17 and 18, | 70° |
| " | " | " | lowest February 28, | 32¾° |
| " | " | " | Range, | 37¼° |
| " | " | " | at four feet depth. | |
| " | " | " | highest July 29, | 64½° |
| " | " | " | lowest February 25, | 35½° |
| " | " | " | Range, | 29° |
| " | " | " | of the air, February 12, | -3° |
| " | " | " | of the air, August, 3, P. M., | 90° |
| " | " | " | Range, | 93° |
| Temperature at Scott | | | at two feet depth. | |
| " | " | " | highest, August 17 and 18, | 64° |
| " | " | " | at four ft. depth, 17 days in Aug. | 60° |
| " | " | " | of the air, at 3, P. M., highest | 90° |

[292] The temperature of falling rain, however, in the hot season, is many degrees cooler than the lower stratum of the atmosphere, and the surface of the earth upon which it falls. The effects of rain on drained soil, in the heat of Summer, are, then, two-fold; to cool the burning surface, which is, as we have seen, much warmer than the rain, and, at the same time, to warm the subsoil which is cooler than the rain itself, as it falls, and very much cooler than the rain-water, as it is warmed by its passage through the hot surface soil. These are beautiful provisions of Nature, by which the excesses of heat and cold are mitigated, and the temperature of the soil rendered more uniform, upon land adapted, by drainage, to her genial influences.

Upon the saturated and water-logged bog, as we have seen, the effect of the greatest heat is insufficient to raise the temperature of the subsoil a single degree, while the surface may be burned up and "shrivelled like a parched scroll."

Drainage also raises the temperature of the soil by the admission of warm air. This proposition is closely connected with that just discussed. When the air is warmer than the soil, as it always is in the Spring-time, the water from the melting snow, or from rain, upon drained land, passes downward, and runs off by its gravitation. As "Nature abhors a vacuum," the little spaces in the soil, from which the water passes, must be filled with air, and this air can only be supplied from the surface, and, being warmer than the ground, tends to raise its temperature. No such effect can be produced in land not drained, because no water runs out of it, and there are, consequently, no such spaces opened for the warm air to enter.

Drainage equalizes the temperature of the soil in Summer by increasing the deposit of dew. Of this we shall speak further, in a future chapter. [293]

*Drainage raises the temperature in Spring by diminishing evaporation.* Evaporation may be defined to be the conversion of liquid and solid bodies into elastic fluids, by the influence of caloric.

By heating water over a fire, bubbles rise from the bottom of the vessel, adhere awhile to the sides of it, and then ascend to the surface, and burst and go off in visible vapor, or, in other words, by evaporation. Water is evaporated by the heat of the sun merely, and even without this heat, in the open air. It is evaporated at very low temperatures, when fully exposed to the air. Even ice evaporates in the open air. We often observe in Winter, that a thin covering of ice or snow disappears from our roads, although there has been no thawing weather.

In another chapter, we have considered the subject of "Evaporation and Filtration," and endeavored to give some general idea of the proportion of the rain which escapes by evaporation. We have seen, that evaporation proceeds much more rapidly from a surface of water, as a pond or river, than from a land surface, unless it be fully saturated, and that evaporation from the water exceeds the whole amount of rain, about as much as evaporation from the land falls short of the amount of rain. Thus, by this simple agency of evaporation, the vast quantities of water that are constantly flowing, in all the rivers of the earth, into the sea, are brought back again to the land, and so the great system of circulation is maintained throughout the ages.

As evaporation is greatest from a water-surface, so it is greater, other things being equal, according to the wetness of the surface of any given field. If the field be covered with water, it becomes a water-surface for the time, and the evaporation is like that from a pond. If, as is often the case, the water stands on it in spots, over half its [294] surface, and the rest is saturated, the evaporation is scarcely less, and has been said to be even more; while, if the surface be comparatively dry, the evaporation is very little.

But what harm does evaporation do? and what has all this scientific talk to do with drainage? These, my friend, are very practical questions, and just the ones which it is proposed to answer; but we must bear in mind that, as Nature conducts her grand affairs by systematic laws, the small portion of her domain which for a brief space of time we occupy, is not exempted from their operation. Some of these laws we may comprehend, and turn our knowledge of them to practical account. Of others, we may note the results, without apprehending the reasons of them; for it is true—

> "There are more things in Heaven and earth, Horatio,
> Than are dreamt of in your philosophy."

Discussions of this kind may seem dry, though the subject itself be moisture. They belong, certainly, to the topic under consideration.

Evaporation does harm in the Spring-time, because it produces cold, just when we most want heat. How it produces cold, is not so readily explained. The fact may be made as evident as the existence of sin in the world, and, possibly, the reason of it may be as unsatisfactory.

The books say, that heat always disappears when a solid body becomes a liquid; and so it is, that the air always remains cool while the snow and ice are melting in Spring. Again, it is said that heat always disappears, when a fluid becomes vapor. These are said to be laws or principles of nature, and are said to explain other phenomena. To a practical mind, it is perhaps just as satisfactory to say that evaporation produces cold, as to state the principle or law in the language of science.

That the fact is so, may be proved by many illustrations. [295] Stockhardt gives the following experiment, which is strikingly appropriate:

"Fill a tube half full of water, and fasten securely round the bulb of it, a piece of cloth. Saturate the cloth with cold water, and then twirl the tube rapidly between the hands; presently the water in the tube will become sensibly colder, and the degree of cold may be accurately determined by the thermometer. Moisten the cloth with ether, a very volatile liquid, and twirl it again in the same manner as before; by which means, its contents, even in Summer, may be converted into ice."

It is very fortunate for us, that our Spring showers are not of ether; for then, instead of thawing, our land would freeze the harder! The heat of the blood is about 98°; yet man can endure a heat of many degrees more, and even labor under a Summer sun, which would raise the thermometer to 130°, without the temperature of his blood being materially affected, and it is because of perspiration, which absorbs the surplus heat, or, in other words, creates cold. It is said, too, that on the same principle, if two saucers, one filled with water warm enough to give off visible vapor, the other filled with water just from the well, are exposed in a sharp frosty morning, that filled with the warm water will exhibit ice soonest. Wine is cooled by evaporation, by wrapping the bottle in wet flannel, and exposing it to the air.

If, after all this, any one doubts the fact that evaporation tends to produce cold, let him countenance his skepticism, by wetting his face with warm water, and going into the air in a Winter's day, and his faith will be greatly strengthened.

We have, in the northern part of America, most water in the soil in the Spring of the year, just at the time when we most need a genial warmth to promote germination. If land is well drained, this water sinks downward, and runs away in the drains, instead of passing upward by evaporation. [296]

Drainage, therefore, diminishes evaporation simply by removing the surplus snow and rain-water by filtration. It thus raises the temperature of the soil in that part of the season, when water is flowing from the drains; but, in the heat of Summer, the influence of the showers which refresh without saturating the soil, and are retained in it by attraction, is not lessened. As a good soil retains by attraction about one-half its weight of water that cannot be drained out, there can be no reasonable apprehension that the "gentle Summer showers" will be wasted by filtration, even upon thorough-drained land, while an avenue is open, by the drains, for the escape of drowning floods.

To show the general effect of drainage, in raising the temperature of wet lands in Summer, the following statement of Mr. Parkes is valuable. An elevation of the temperature of the subsoil ten degrees, will be seen to be very material, when we consider that Indian corn will not vegetate at all at 53°, but will start at once at 63°, 55° being its lowest point of germination:

"As regards the temperature of the water derived from drainage at different seasons of the year, I am unacquainted with any published facts. This is a subject of the highest import, as thermometric observations may be rendered demonstrative, in the truest manner, of the effect of drainage on the climate of the soil. At present, I must limit myself to saying, that I have never known the water of drainage issue from land drained at Midsummer, to depths of four and five feet, at a higher temperature than 52° or 53° Fahrenheit: whereas, in the following year and subsequent years, the water discharged from the same drains, at the same period, will issue at a temperature of 60°, and even so high as 63°, thus exhibiting the increase of heat conferred during the Summer months on the terrestrial climate by drainage. This is the all-important fact connected with the art and science of land-drainage."

Besides affecting favorably the temperature of the particular field which is drained, the general effect of the drainage of wet lands upon the climate of the neighborhood has often been noticed. In the paper already cited, emanating from the Board of Health, we find the following [297] remarks, which are in accordance with all observation in districts where under-drainage has been generally practiced:

"Every one must have remarked, on passing from a district with a retentive soil to one of an open porous nature—respectively characterized as cold and warm soils—that, often, whilst the air on the retentive soil is cold and raw, that on the drier soil is comparatively warm and genial. The same effect which is here caused naturally, may be produced artificially, by providing for the perfect escape of superfluous water by drainage, so as to leave less to cool down the air by evaporation. The reason of this difference is two-fold. In the first place, much heat is saved, as much heat being required for the vaporization of water, as would elevate the temperature of more than three million times its bulk of air one degree. It follows, therefore, that for every inch in depth of water carried off by drains, which must otherwise evaporate, as much heat is saved per acre as would elevate eleven thousand million cubic feet of air one degree in temperature. But that is not all. Not only is the temperature of the air reduced, but its dew point is raised, by water being evaporated which might be drained off; consequently, the want of drainage renders the air both colder and more liable to the formation of dew and mists, and its dampness affects comfort even more than its temperature. It is easy, then, to understand how local climate is so much affected by surplus moisture, and so remarkably improved by drainage. A farmer being asked the effect on temperature of some new drainage works; replied, that all he knew was, that before the drainage he could never go out at night without a great coat, and that now he could, so that he considered it made the difference of a great coat to him."

*Drainage increases the coldness of the subsoil in Winter.* Whether this is a gain or loss to the agriculturist, is not for us to determine. The object of our labor is, to lay the whole subject fairly before the reader, and not to extol drainage as the grand panacea of bad husbandry.

Although water will not conduct heat downwards, yet it doubtless prevents the deep freezing of the ground. It has already been seen, that the temperature of the earth, a few feet below the surface, is above the freezing point, at all times. The fact that the ground does not freeze, usually, even in New England, where every Winter brings [298] weather below Zero, more than four or five feet deep, in the most exposed situations, shows conclusively the comparatively even temperature of the subsoil. The water which flows underground is of this subsoil temperature, and, in Winter, warms the ground through which it flows. In land thoroughly drained, this warm water cannot rise above the drains, and so cannot defend the soil from frost.

Drained land will, undoubtedly, freeze deeper than undrained land, and this is a fact to be impressed upon all who lay tiles in a cold climate. It is a strong argument for deep drainage. "Drain deep, or drain not," is a convenient paraphrase of a familiar quotation. How often do we hear it said, "My meadow never freezes more than a foot deep; there will never be any trouble from frost in that place, if the tiles are no more than two feet deep." Be assured, brother farmer, that the frost will follow the water-table downward, and, unless the warm water move in sufficient quantity through your pipes to protect them in Winter, your work may be ruined by frost. So long as much water is flowing in pipes, especially if it be from deep springs, they will be safe from frost, even at a slight depth.

Dr. Madden says, that it has been proved that one great source of health and vigor in vegetation, is the great difference which exists between the temperature of Summer and Winter, which, he says, in dry soils, often amounts to between 30° and 40°; while, in very wet soils, it seldom exceeds 10°. This idea may have value in a mild climate; but, probably, in New England, we get cold enough for our good, without artificial aids. In another view, drainage is known to be essential, even in Winter.

Fruit trees are almost as surely destroyed by standing with their feet in cold water all Winter, as any of us "unfeathered bipeds" would be; while the solid freezing [299] of the earth around their roots does not harm them. Perhaps the same is true of most other vegetation.

The deep freezing of the ground is often mentioned as a mode of pulverization—as a sort of natural subsoiling thrown in by a kind Providence, by way of compensation for some of the evils of a cold climate. Most of those, however, who have wielded the pick-axe in laying four-foot drains, in clay or hard-pan, will have doubts whether Jack Frost, though he can pull up our fence-posts, and throw out our Winter grain, has much softened the earth two feet below its surface.

That the frost comes out of drained land earlier than undrained, in Spring, we are satisfied, both by personal observation, and by the statements of the few individuals who have practiced thorough-drainage in our cold climate.

B. F. Nourse, Esq., whose valuable statement will be found in a later chapter, says, that, in 1858, the frost came out a week, at least, earlier from his drained land, in Maine, than from contiguous un-drained land; and that, usually, the drained land is in condition to be worked as soon as the frost is out, quite two weeks earlier than any other land in the vicinity. Our observations on our own land, fully corroborate the opinion of Mr. Nourse.

The reasons why the frost should come out of drained land soon-est, are, that land that is dry does not freeze so solid as land that is wet, and so spaces are left for the permeation of warm air. Again, ice, like water, is almost a nonconductor of heat, and earth saturated with water and frozen, is like unto it, so that neither the warmth of the subsoil or surface-soil can be readily imparted to it. Dry earth, on the other hand, although frozen, is still a good conductor, and readily dissolves at the first warm breath of Spring above, or the pulsations of the great heart of Nature beneath.

# CHAPTER XVI. [300]
# POWER OF SOILS TO ABSORB AND RETAIN MOISTURE.

Why does not Drainage make the Land too Dry? — Adhesive Attraction. — The Finest Soils exert most Attraction. — How much Water different Soils hold by Attraction. — Capillary Attraction, Illustrated. — Power to Imbibe Moisture from the Air. — Weight Absorbed by 1,000 lbs. in 12 Hours. — Dew, Cause of. — Dew Point. — Cause of Frost. — Why Covering Plants Protects from Frost. — Dew Imparts Warmth. — Idea that the Moon Promotes Putrefaction. — Quantity of Dew.

The first and most natural objection made, by those not practically familiar with drainage operations, to the whole system is, that the drains will draw out so much of the water from the soil, as to leave it too dry for the crops.

If a cask be filled with round stones, or with musket balls, or with large shot, and with water to the surface, and then an opening be made at the bottom of the cask, all the water, except a thin film adhering to the surface of the vessel and its contents, will immediately run out.

If now, the same cask be filled with the dried soil of any cultivated field, and this soil be saturated with water, a part only of the water can be drawn out at the bottom. The soil in the cask will remain moist, retaining more or less of the water, according to the character of the soil.

Why does not the water all run out of the soil, and leave it dry? An answer may be found in the books, which is, in reality, but a restatement of the fact, by reference to a principle of nature, by no means intelligible to finite minds, called attraction. If two substances are [301] placed in close contact with each other, they cannot be separated without a certain amount of force.

"If we wet the surfaces of two pieces of glass, and place them in contact, we shall find that they adhere to each other, and that, independently of the effect of the pressure of the air, they oppose considerable resistance to any attempt to separate them. Again, if we bring any substance, as the blade of a knife, in contact with water, the water adheres to the blade in a thin film, and remains, by what is termed *adhesive attraction*. This property resides in the surface of bodies, and is in proportion to the extent of its surface.

"Soils possess this property, in common with all other bodies, and possess it, in a greater or less degree, according to the aggregate surface which the particles of a given bulk present. Thus, clay may, by means of kneading, be made to contain so large a quantity of water, as that, at last, it may almost be supposed to be divided into infinitesimally thin layers, having each a film of water adhering to it on either side. Such soils, again, as sand or chalk, the particles of which are coarser exert a less degree of adhesive attraction for water."—*Cyc. of Ag.*, 695.

Professor Schübler, of Tubingen, gives the results of experiments upon this point. By dropping water upon dried soils of different kinds, until it began to drop from the bottom, he found that 100 lbs. of soil held by attraction, as follows:

| | | |
|---|---|---|
| Sand | 25 | lbs. of water. |
| Loamy Soil | 40 | " |
| Clay Loam | 50 | " |
| Pure Clay | 70 | " |

Mr. Shedd, of Boston, gives the result of a recent experiment of his own on this point. He writes thus:

"I have made an experiment with a soil of ordinary tenacity, to ascertain how much water it would hold in suspension, with the following result: One cubic foot of earth held 0.4826434 cubic feet of water; three feet of dry soil of that character will receive 1.44793 ft. vertical depth of water before any drains off, or seventeen and three-quarter inches, equal to nearly six month's rain-fall. One cubic foot of earth held 3.53713 gallons of water, or if drains are three feet deep, one square foot of surface would receive 10.61 gallons of water, before [302] saturation. Other soils would sustain a greater or less quantity, according to their character."

Besides this power of retaining water, when brought into contact with it, the soil has, in common with other porous bodies, the power of drawing up moisture, or of absorbing it, independent of gravitation, or of the weight of the water which aids to carry it down into the soil. This power is called *capillary attraction*, from the hair-like tubes used in early experiments. If very minute tubes, open at both ends, are placed upright, partly immersed in a vessel of water, the water rises in the tubes perceptibly higher than its general surface in the vessel. A sponge, from which water has been pressed out, held over a basin of water, so that its lower part touches the surface, draws up the water till it is saturated. A common flower-pot, with a perforated bottom, and filled with dry earth, placed in a saucer of water, best illustrates this point. The water rises at once to a common level in the pot and outside. This represents the water-table in the soil of our fields. But, from this level, water will continue to rise in the earth in the pot, till it is moistened to the surface, and this, too, is by capillary attraction.

The tendency of water to ascend, however, is not the same in all soils. In coarse gravelly soils, the principle may not operate perfectly, because the interstices are too large, the weight of the water overcoming the power of attraction, as in the cask of stones or shot. In very fine clay, on the other hand, although it be absorptive and retentive of water, yet the particles are so fine, and the spaces between them so small, that this attraction, though sure, would be slow in operation. A loamy, light, well pulverized soil, again, would perhaps furnish the best medium for the diffusion of water in this way.

It is impossible to set limits to so uncertain a power as this of capillary attraction. We see that in minute glass [303] tubes, it has power to raise water a small fraction of an inch only. We see that, in the sponge or flower-pot, it has power to raise water many inches; and we know that, in the soil, moisture is thus attracted upwards several feet. By observing a saturated sponge in a saucer, we shall see that, although moist at the top, it holds more and more water to the bottom. So, in the saturated earth in a flower-pot, the earth, merely moist at the surface, is wet mud just above the water-table. So, in drained land, the capillary force which retained the water in the soil to the height of a few inches, is no longer able to sustain it, when the height is increased to feet, and a portion descends into the drain, leaving the surface comparatively dry.

Thus, it would seem, that draining may modify the force of capillary attraction, while it cannot affect that of adhesive attraction. It may drain off surplus water, but, unaided, can never render any arable land too dry. If, however, the surplus water be speedily taken off by drainage, and the capillary attraction be greatly impaired, so that little water is drawn upwards by its force, will not the soil soon become parched by the heat of the sun, or, in other words, by evaporation?

Without stopping in this place, to speak of evaporation, we may answer, that, in our burning Summer heat, the earth would be burnt up too dry for any vegetation, were it not for a beneficent arrangement of Providence, which counteracts the effect of the sun's rays, and of which we will now make mention.

*Power to imbibe moisture from the air.*—We have spoken, in another place, of the absorption, by drained land, of fertilizing substances from the atmosphere. Dry soil has, too, a wonderful power of deriving moisture from the same source.

"When a portion of soil," says Johnston, "is dried carefully over boiling water, or in an oven, and is then spread out upon a sheet of [304] paper in the open air, it will gradually drink in watery vapor from the atmosphere, and will thus increase in weight.

"In hot climates and in dry seasons, this property is of great importance, restoring as it does, to the thirsty soil, and bringing within the reach of plants, a portion of the moisture, which, during the day, they had so copiously exhaled."

Different soils possess this power in unequal degrees. During a night of 12 hours, and when the air is moist, according to Schübler, 1000 lbs. of perfectly dry

Quartz sand will gain                                              0     lbs.

| | | |
|---|---|---|
| Calcareous sand | 2 | " |
| Loamy soil | 21 | " |
| Clay loam | 25 | " |
| Pure agricultural clay | 27 | " |

Sir Humphrey Davy found, that the power of attraction for water, generally proved an index to the agricultural value of soils. It is, however, but one means of judging of their value. Peaty soils and strong clays are very absorbent of water, although not always the best for cultivation.

Sir H. Davy gives the following results of his experiments. When made perfectly dry, 1000 lbs. of a

| | | |
|---|---|---|
| Very fertile soil from East Lothian, gained in an hour | 18 | lbs. |
| Very fertile soil from Somersetshire | 16 | " |
| Soil, worth 45s., (rent) from Essex | 13 | " |
| Sandy soil, worth 28s., from Essex | 11 | " |
| Coarse sand, worth 15s. | 8 | " |
| Soil of Bagshot Heath | 3 | " |

"This sort of attraction, however," suggests a writer in the Cyclopedia of Agriculture, "it may be believed, depends upon other causes besides the attraction of adhesion. The power of attraction, which certain substances exhibit for the *vapor* of water, is more akin to the force which enables certain porous bodies to absorb and retain many times their volume of the different gases; as charcoal, of ammonia, of which it is said to absorb ninety times its own bulk."

Here again, we find in the soil, an inexplicable but beneficent [305] power, by which it supplies itself with moisture when it most needs it.

Warm air is capable of holding more vapor than cooler air, and the very heat of Summer supplies it with moisture by evaporation from land and water. As the air is cooled, at nightfall, it must somewhere deposit the water, which the hand of the Unseen presses out of it by condensation.

The sun-dried surface of fertile, well drained soil, is in precisely the condition best adapted to receive the refreshing draught, and convey it to the thirsting plants.

We may form some estimate of the vast amount absorbed by an acre of land in a dry season, by considering that the clay loam, in the above statement, absorbed in 12 hours a fortieth part of its own weight.

## OF DEW.

Dew is one of the most ordinary forms in which moisture is deposited in and upon the soil, in its natural conditions. The absorbent power of artificially-dried soils, as has been seen, seems to depend much upon their chemical constitution; and that topic has been considered, without special reference to the comparative temperature of the soil and atmosphere. The soil, as we have seen, absorbs moisture from the air, when both are of the same temperature, the amount absorbed depending also upon the physical condition of the soil, and upon the comparative moisture of the soil and atmosphere.

The deposition of dew results from a different law. All bodies throw off, at all times, heat, by radiation, as it is termed. In the daytime, the sun's rays warm the earth, and the air is heated by it, and that nearest the surface is heated most. Evaporation is constantly going on from the earth and water, and loads the air with vapor, and the warmer the air, the more vapor it will hold.

When the sun goes down, the earth still continues to [306] throw off heat by radiation, and soon becomes cooler than the air, unless the same amount of heat be returned, by radiation from other surfaces. Becoming cooler than the air, the soil or plants cool the air which comes in contact with them; and thus cooled to a certain point, the air cannot hold all the vapor which it absorbed while warmer, and part of it is deposited upon the soil, plant, or other cool surface. This is dew; and the temperature at which the air is saturated with vapor, is called the dew-point. If saturated at a given temperature with vapor, the air, when cooled below this point, must part with a portion of the vapor, in some way; in the form of rain or mist, if in the air; in the form of dew, if on the surface of the earth.

If, however, other surfaces, at night, radiate as much heat back to the earth as it throws off, the surface of the earth is not thus cooled, and there is no dew. Clouds radiate heat to the earth, and, therefore, there is less dew in cloudy than in clear nights. If the temperature of the earth sinks below the freezing-point, the aqueous vapor is frozen, and is then called *frost*.

To radiate back a portion of the heat thus thrown off by the soil and plants, gardeners cover their tender plants and vines with mats or boards, or even with thin cloth, and thus protect them from frost. If the covering touch the plants, they are often frozen, the heat being conducted off, by contact, to the covering, and thence radiated. Dew then is an effect, but not a cause, of cold. It imparts warmth, because it can be deposited only on objects cooler than itself.

It has been supposed by many that the light of the moon promotes putrefaction. Pliny and Plutarch both affirm this to be true. Dew, by supplying moisture in the warm season, aids this process of decay. We have seen that dew is most abundant in clear nights; and although all clear nights are not moonlight nights, yet all moonlight [307] nights are clear nights; and this, perhaps, furnishes sufficient grounds for this belief, as to the influence of the moon.

The quantity of dew deposited is not easily measured. It has, however, been estimated by Dr. Dalton, to amount, in England, to five inches of water in a year, or 500 tons to the acre, equal to about one quarter of our rain-fall during the six summer months!

Deep and well-pulverized soils attract much more moisture, in every form, from the atmosphere, than shallow and compact soils. They, in fact, expose a much larger surface to the air. This is the reason why stirring the ground, even in the Summer drought, refreshes our fields of Indian corn.

# CHAPTER XVII [308]
# INJURY OF LAND BY DRAINAGE.

Most Land cannot be Over-drained. — Nature a Deep drainer. — Over-draining of Peaty Soils. — Lincolnshire Fens; Visit to them in 1857. — 56 Bushels of Wheat to the Acre. — Wet Meadows subside by Drainage. — Conclusions.

Is there no danger of draining land too much? May not land be over-drained? These are questions often and very naturally asked, and which deserve careful consideration. The general answer would be that there is no danger to be apprehended from over-draining; that no water will run out of land that would be of advantage to our cultivated crops by being retained. In other words, soils *generally* hold, by capillary attraction, all the moisture that is of any advantage to the crops cultivated on them; and the water of drainage would, if retained for want of outlets, be stagnant, and produce more evil than good.

We say this is generally true; but there are said to be exceptional cases, which it is proposed to consider. If we bear in mind the condition of most soils in Summer, we shall see that this apprehension of over-draining is groundless. The fear is, that crops will suffer in time of drought, if thoroughly drained. Now, we know that, in almost all New England, the water-table is many feet below the surface. Our wells indicate pretty accurately where the water-table is, and drains, unless cut as low as the surface [309] of the water in the wells, would not run a drop of water in Summer.

Our farmers dig their wells twenty, and even fifty, feet deep, and expect that, every Summer, the water will sink to nearly that depth; but they have no apprehension that their crops will become dry, because the water is not kept up to within three feet of the surface.

The fact is, that Nature drains thoroughly the greater portion of all our lands; so that artificial drainage, though it may remove surplus water from them more speedily in Spring, cannot make them more dry in Summer. And what thus happens naturally, on most of the land, without injury, cannot be a dangerous result to effect by drainage on lands of similar character. By thorough-drainage, we endeavor to make lands which have an impervious or very retentive subsoil near the surface, sufficiently open to allow the surplus water to pass off, as it does naturally on our most productive upland.

## OVER-DRAINING OF PEATY SOILS.

No instance has yet been made public in America, of the injury of peat lands by over-drainage; but there is a general impression among English writers, that peat soils are often injured in this way. The Lincolnshire Fens are cited by them, as illustrations of the fact, that these lands do not require deep drainage.

Mr. Pusey says, "Every one who is practically acquainted with moory land, knows that such land may be easily over-drained, so that the soil becomes dusty or *husky*, as it is called—that is, like a dry sponge—the white crops flag, and the turnip leaves turn yellow in a long drought."

These Fens contain an immense extent of land. The Great Level of the Fens, it is said, contains 600,000 acres. [310] Much of this was formerly covered by the tides, and all of it, as the name indicates, was of a marshy character. The water being excluded by embankments against the sea and rivers, and pumped out by steam engines, and the land under-drained generally with tiles, so that the height of the water is under the control of the proprietors, grave disputes have arisen as to the proper amount of drainage.

An impression has heretofore prevailed, that these lands would be too dry if the water were pumped out, so as to reduce the water-table more than a foot or two below the surface, but this idea is now controverted.

In July 1857, in company with three of the best farmers in Lincolnshire, the writer visited the Fens, and carefully examined the crops and drainage. We passed a day with one of the proprietors, who gave us some information upon the point in question. He stated, that in general, the occupants of this land entertain the opinion, that the crops would be ruined by draining to the depth of four feet. So strongly was he impressed with the belief that a deeper drainage was desirable, that he had enclosed his own estate with separate embankments, and put up a steam-engine, and pumped out the water to the depth of four feet, while from the land all around him, it is pumped out only a foot and a half below the surface, though in Summer it may sometimes fall somewhat lower.

The crops on this land were astonishing. Our friends estimated that the wheat then growing and nearly ripe, would yield fifty-six bushels to the acre. Although this was considered a very dry season, the crops on the land of our host were fully equal to the best upon the Fens.

The soil upon that part of the Fens is now a fine black loam of twelve or eighteen inches depth, resting upon clay. Upon other portions, the soil is of various depth and character, resting sometimes upon gravel. [311]

Attention is called to these facts here, to show that the common impression that these lands will not bear deep drainage, is controverted among the occupants themselves, and may prove to be one of those errors which becomes traditional, we hardly know how.

Most peat meadows, in New England, when first relieved of stagnant water, are very light and spongy. The soil is filled with acids which require to be neutralized by an application of lime, or what is cheaper and equally effectual, by exposure to the atmosphere. These soils, when the water is suddenly drawn out of them, retain their bulk for a time, and are too porous and unsubstantial for cultivation. A season or two will cure this evil, in many cases. The soil will become more compact, and will often settle down many inches. It is necessary to bear this in mind in adjusting the drains, because a four-foot drain, when laid, may, by the mere subsidence of the land, become a three-foot drain.

A hasty judgment, in any case, that the land is over-drained, should be suspended until the soil has acquired compactness by its own weight, and by the ameliorating effect of culture and the elements.

Mr. Denton, alluding to the opinion of "many intelligent men, that low meadow-land should be treated differently to upland pasture, and upland pasture differently to arable land," says, "My own observations bring me to the conclusion, that it is not possible to lay pasture-land too dry; for I have invariably remarked, during the recent dry Summer and Autumn particularly, that both in lowland meadows, and upland pastures, those lands which have been most thoroughly drained by deep and frequent drains, are those that have preserved the freshest and most profitable herbage."

While, therefore, we have much doubt whether any land, high or low, can be over-drained for general cultivation, [312] it is probable that a less expensive mode of drainage may be sometimes expedient for grass alone.

While we believe that, in general, even peat soils may be safely drained to the same depth with other soil, there seems to be a well-founded opinion that they may frequently be rendered productive by a less thorough system.

The only safety for us, is in careful experiment with our own lands, which vary so much in character and location, that no precise rules can be prescribed for their treatment.

# CHAPTER XVIII [313]
# OBSTRUCTION OF DRAINS.

Tiles will fill up, unless well laid.—Obstruction by Sand or Silt.—Obstructions at the Outlet from Frogs, Moles, Action of Frost, and Cattle.—Obstruction by Roots.—Willow, Ash, &c., Trees capricious.—Roots enter Perennial Streams.—Obstruction by Mangold Wurtzel.—Obstruction by Per-Oxide of Iron.—How Prevented—Obstruction by the Joints Filling.—No Danger with Two-Inch Pipes.—Water through the Pores.—Collars.—How to Detect Obstructions.

But won't these tiles get filled up and stopped? asks almost every inquirer on the subject of tile draining.

Certainly, they will, if not laid with great care, and with all proper precautions against obstructions. It cannot be too often repeated, that tile-drainage requires science, and knowledge, and skill, as well as money; and no man should go into it blindfold, or with faith in his innate perceptions of right. If he does, his education will be expensive.

It is proposed to mention all the various modes by which tiles have been known to be obstructed, and to suggest how the danger of failure, by means of them, may be obviated.

Let not enterprising readers be alarmed at such an array of difficulties, for the more conspicuous they become, the less is the danger from them.

*Obstruction by Sand or Silt.* Probably, more drains are rendered worthless, by being filled up with earthy matter, which passes with water through the joints of the tiles, than by every other cause. [314]

Fine sand will pass through the smallest aperture, if there is a current of water sufficient to move it, and silt, or the fine deposit of mud or other earth, which is held almost in solution in running water, is even more insinuating in its ways than sand.

Very often, drains are filled up and ruined by these deposits; and, unless the fall be considerable, and the drain be laid with even descent, if earth of any kind find entrance, it must endanger the permanency of the work. To guard against the admission of everything but water, lay drains deep enough to be beyond the danger of water bursting in, in streamlets. Water should enter the drain at the bottom, by rising to the level of the tiles, and not by sinking from the surface directly to them. If the land is sandy, great care must be used. In draining through flowing sand, especially if there be a quick descent, the precaution of sheathing tiles is resorted to. That is done by putting small tiles inside of larger ones, breaking joints inside, and thus laying a double drain. This is only necessary, however, in spots of sand full of spring-water. Next best to this mode, is the use of collars over the joints, but these are not often used, though recommended for sandy land.

At least, in all land not perfectly sound, be careful to secure the joints in some way. An inverted turf, carefully laid over the joint, is oftenest used. Good, clean, fine gravel is, perhaps, best of all. Spent tan bark, when it is to be conveniently procured, is excellent, because it strains out the earth, while it freely admits water; and any particles of tan that find entrance, are floated out upon the water. The same may be said of sawdust.

To secure the exit of earth that may enter at the joints, there should be care that the tiles be smooth inside, that they be laid exactly in line, and that there be a continuous descent. If there be any place where the water rises in the [315] tiles, in that place, every particle of sand, or other matter heavier than water, will be likely to stop, until a barrier is formed, and the drain stopped.

In speaking of the forms of tiles, the superiority of rounded openings over those with flat bottom has been shown. The greater head of water in a round pipe, gives it force to drive before it all obstructions, and so tends to keep the drain clear.

*Obstructions at the Outlet.* The water from deep drains is usually very clear, and cattle find the outlet a convenient place to drink at, and constantly tread up the soft ground there, and obstruct the flow of water. All earthy matter, and chemical solutions of iron, and the like, tend to accumulate by deposit at the outlet. Frogs and mice, and insects of many kinds, collect about such places, and creep into the drains. The action of frost in cold regions displaces the earth, and even masonry, if not well laid; and back-water, by flowing into the drains, hinders the free passage of water.

All these causes tend to obstruct drains at the outlet. If once stopped there, the whole pipe becomes filled with stagnant water, which deposits all its earthy matter, and soon becomes obstructed at other points, and so becomes useless. The outlet must be rendered secure from all these dangers, at all seasons, by some such means as are suggested in the chapter on the Arrangement of Drains.

*Obstruction by roots.* On the author's farm in Exeter, a wooden drain, to carry off waste water from a watering place, was laid, with a triangular opening of about four inches. This was found to be obstructed the second year after it was laid; and upon taking it up, it proved to be entirely filled for several feet, with willow roots, which grew like long, fine grass, thickly matted together, so as entirely to close the drain. There was a row of large willows about thirty feet distant, and as the drain was but [316] about two feet deep, they found their way easily to it, and entering between the rough joints of the boards, not very carefully fitted, fattened on the spring water till they outgrew their new house.

A neighbor says, he never wants a tree within ten rods of any land he desires to plow; and it would be unsafe to undertake to set limits to the extent of the roots of trees. "No crevice, however small," says a writer, "is proof against the entrance of the roots of water-loving trees."

The behavior of roots is, however, very capricious in this matter; for, while occasional instances occur of drains being obstructed by them, it is a very common thing for drains to operate perfectly for indefinite periods, where they run through forests and orchards for long distances. They, however, who lay drains near to willows and ashes, and the like cold-water drinkers, must do it at the peril of which they are warned.

Laying the tiles deep and with collars will afford the best security from all danger of this kind.

Thos. Gisborne, Esq., in a note to the edition of his Essay on Drainage published in 1852, says:

My own experience as to roots, in connection with deep pipe draining, is as follows: — I have never known roots to obstruct a pipe through which there was not a perennial stream. The flow of water in Summer and early Autumn appears to furnish the attraction. I have never discovered that the roots of any esculent vegetable have obstructed a pipe. The trees which, by my own personal observation, I have found to be most dangerous, have been red willow, black Italian poplar, alder, ash, and broad-leaved elm. I have many alders in close contiguity with important drains; and, though I have never convicted one, I can not doubt that they are dangerous. Oak, and black and white thorns, I have not detected, nor do I suspect them. The guilty trees have, in every instance, been young and free growing; I have never convicted an adult.

Mangold-wurzel, it is said by several writers, will sometimes grow down into tile drains, even to the depth of four feet, and entirely obstruct them; but those are cases of [317] very rare occurrence. In thousands of instances, mangolds have been cultivated on drained land, even where tiles were but 2½ feet deep, without causing any obstruction of the drains. Any reader who is curious in such matters, may find in the appendix to the 10th Vol. of the Journal of the Royal Ag. Soc., a singular instance of obstruction of drains by the roots of the mangold, as well as instances of obstructions by the roots of trees.

*Obstruction by Per-oxide of Iron.* In the author's barn-cellar is a watering place, supplied by a half-inch lead pipe, from a spring some eight rods distant. This pipe several times in a year, sometimes once a week, in cold weather, is entirely stopped. The stream of water is never much larger than a lead pencil. We usually start it with a sort of syringe, by forcing into the outlet a quantity of water. It then runs very thick, and of the color of iron rust, sometimes several pails full, and will then run clear for weeks or months, perhaps. In the tub which receives the water, there is always a large deposit of this same colored substance; and along the street near by, where the water oozes out of the bank, there is this same appearance of iron. This deposit is, in common language, called per-oxide of iron, though this term is not, by chemists of the present day, deemed sufficiently accurate, and the word sesqui-oxide is preferred in scientific works.

Iron exists in all animal and vegetable matter, and in all soils, to some extent. It exists as protoxide of iron, in which one atom of iron always combines with one atom of oxygen, and it exists as sesqui-oxide of iron, from the Latin *sesqui*, which means one and a half, in which one and a half atoms of oxygen combine with one atom of iron. The less accurate term, per-oxide, has been adopted here, because it is found in general use by writers on drainage.

The theory is that the iron exists in the soil, and is held [318] in solution in water as a protoxide, and is converted into per-oxide by contact with the air, either in the drains or at their outlets, and is then deposited at the bottom of the water.

In a pipe running full there would be, upon this theory, no exposure to the air, which should form the per-oxide. In the case stated, it is probable that the per-oxide is formed at the exposed surface of a large cask, at the spring, and is carried into the pipe, as it is precipitated. Common drain pipes would be full of air, which might, perhaps, in a feeble current, be sufficient to cause this deposit.

Occasionally, cases have occurred of obstruction from this cause, and whenever the signs of this deposit are visible about the field to be drained, care must be used to guard against it in draining.

To guard against obstruction from per-oxide of iron, tiles should be laid deep, closely jointed or collared, with great care that the fall be continuous, and especially that there be a quick fall at the junctions of minor drains with mains, and a clear outlet.

Mr. Beattie, of Aberdeen, says: Before adopting 4 feet drains, I had much difficulty in dealing with the iron ore which generally appeared at two to three feet from the surface, but by the extra depth the water filters off to the pipes free of ore. Occasionally, iron ore is found at a greater depth, but the floating substance is then in most cases lighter, and does not adhere to the pipes in the same way as that found near the surface. Arrangements should also be made for examining the drains by means of wells, and for flushing them by holding back the water until the drains are filled, and then letting it suddenly off, or, by occasionally admitting a stream of water at the upper end, when practicable, and thus washing out the pipes. Mr. Denton says: "It is found that the use of this contrivance [319] for flushing, will get rid of the per-oxide of iron, about which so much complaint is made."

*Obstruction by Filling at the Joints.* One would suppose that tiles might frequently be prevented from receiving water, by the filling up of the crevices between them. If water poured on to tiles in a stream, it would be likely to carry into these openings enough earthy matter to fill them; but the whole theory of thorough-drainage rests upon the idea of slow percolation—of the passage of water in the form of fine dew, as it were—through the motionless particles which compose the soil; and, if drains are properly laid, there can be no motion of particles of earth, either into or towards the tiles. The water should soak through the ground precisely as it does through a wet cloth.

In an article in the Journal of the Society of Arts, published in 1855, Mr. Thomas Arkell states that in 1846 he had drained a few acres with 1¼ inch pipes, about three feet deep, and 21 to 25 feet apart. The drains acted well, and the land was tolerably dry and healthy for the first few years; but afterwards, in wet seasons, it was very wet, and appeared full of water, like undrained land, although at the time all the drains were running, but very slowly. His conclusion was that mud had entered the crevices, and stopped the water out. He says he has known other persons, who had used small pipes, who had suffered in the same way. There are many persons still in England, who are so apprehensive on this point, that they continue to use horse-shoe tiles, or, as they are sometimes called, "tops and bottoms," which admit water more freely along the joints.

The most skillful engineers, however, decidedly prefer round pipes, but recommend that none smaller than one-and-a-half-inch be used, and prefer two-inch to any smaller size. The circumference of a two-inch pipe is not far [320] from nine inches, while that of a one-inch pipe, of common thickness, is about half that, so that the opening is twice as extensive in the two-inch, pipes as in the one-inch pipe.

The ascertained instances of the obstruction of pipes, by excluding the water from the joints, are very few. No doubt that clay, puddled in upon the tiles when laid, might have this effect; but they who have experience in tile-drainage, will bear witness that there is far more difficulty in excluding sand and mud, than there is in admitting water.

It is thought, by some persons, that sufficient water to drain land may be admitted through the pores of the tiles. We have no such faith. The opinion of Mr. Parkes, that about 500 times as much water enters at the crevices between each pair of tiles, as is absorbed through the tiles themselves, we think to be far nearer the truth.

Collars have a great tendency to prevent the closing up of the crevices between tiles; but injuries to drains laid at proper depths, with two-inch pipes, even without collars, must be very rare. Indeed, no single case of a drain obstructed in this way, when laid four feet deep, has yet come within our reading or observation, and it is rather as a possible, than even a probable, cause of failure, that it has been mentioned.

## HOW TO DETECT OBSTRUCTIONS IN DRAINS.

When a drain is entirely obstructed, if there is a considerable flow of water, and the ground is much descending, the water will at once press through the joints of the pipes, and show itself at the surface. By thrusting down a bar along the course of the drain, the place of the obstruction will be readily determined; for the water will, at the point of greatest pressure, burst up in the hole made by the bar, like a spring, while below the point of obstruction, [321] there will be no upward pressure of the water, and above it, the pressure will be less the farther we go.

The point being determined, it is the work of but few minutes to dig down upon the drain, remove carefully a few pipes, and take out the frog, or mouse, or the broken tile, if such be the cause of the difficulty. If silt or earth has caused the obstruction, it is probably because of a depression in the line of the drain, or a defect in some junction with other drains, and this may require the taking up of more or less of the pipes.

If there be but little fall in the drains, the obstruction will not be so readily found; but the effect of the water will soon be observed at the surface, both in keeping the soil wet, and in chilling the vegetation upon it. If proper peep-holes have been provided, the place of any obstruction may readily be determined, at a glance into them.

Upon our own land, we have had two or three instances of obstruction by sand, very soon after the tiles were laid, and always at the junction of drains imperfectly secured with bricks, before we had procured proper branch-pipes for the purpose.

A little experience will enable the proprietor at once to detect any failure of his drains, and to apply the proper remedy. Obstructions from silt and sand are much more likely to occur during the first season after the drains are laid, than afterwards, because the earth is loose about the pipes, and more liable to be washed into the joints, than after it has become compact.

On the whole, we believe the danger to tile-drains, of obstruction, is very little, provided good tiles are used, and proper care is exercised in laying them.

# CHAPTER XIX [322]
# DRAINAGE OF STIFF CLAYS.

Clay not impervious, or it could not be wet and dried.—
Puddling, what is.—Water will stand over Drains on Puddled
Soil.—Cracking of Clays by Drying.—Drained Clays improve by
time.—Passage of Water through Clay makes it permeable.—
Experiment by Mr. Pettibone, of Vermont.—Pressure of Water in
saturated Soil.

It is a common impression that clay is impervious to water, and
that, therefore, a clay soil cannot be drained, especially by deep
under-drains. A moment's reflection will satisfy any one that such
land is not absolutely impervious. We find such land is wet in
Spring, at any depth; and, in the latter part of Summer, we find it
comparatively dry. How comes it wet, at any time, if water does not
go into it? And how comes it dry, at any time, if water does not
come out of it?

In treating of the power of the soil to absorb moisture, we have
shown that a clay soil will absorb more than half its weight and
bulk of water, and that it holds more water than any other soil,
with, perhaps, the single exception of peat.

The facts, however, that clay may be wet, and may be dried, and
that it readily absorbs large quantities of water, though they prove
conclusively that it is not impervious to water, yet do not prove that
water will pass through it with sufficient rapidity to answer the
practical purposes of drainage for agriculture. This point can only
be satisfactorily determined by experiment. It is not necessary,
however, that each farmer should try the experiment [323] for himself; because, although we are very apt to think our own case an
exception to all general rules, it is not really probable that any new
kind of clay will be discovered hereafter, that is so different from all
other clay that is known, that established principles will not apply
to it. So far as our own observation extends, owners of clay farms
always over-estimate the difficulty of draining their land. There are
certain notorious facts with regard to clay, which mislead the judgment of men on this point. One of these facts is, that clay is used for

stopping water, by the process called *puddling*. Puddled clay is used for the bottom of ponds, and of canals, and of reservoirs, and, for such purposes, is regarded as nearly, or quite impervious.

We see that, on our clay fields, water stands upon the surface, especially in the ruts of wheels, and on headlands much trodden, late in the season, and when, in other places, it has disappeared. This is due, also, to puddling.

Puddling is merely the working of wet clay, or other soil, by beating, or treading, or stirring, until its particles are so finely divided that water has an exceedingly slow passage between them, with ordinary pressure. We see the effect of this operation on common highways, where water often stands for many days in puddles, because the surface has been ground so fine, and rendered so compact, by wheels and horses, that the water cannot find passage. This, however, is not the natural condition of any clay; nor can any clay be kept in this condition, except by being constantly wet. If once dried, or subjected to the action of frost, the soil resumes its natural condition of porosity, as will be presently explained. They who object to deep drainage, or to the possibility of draining stiff clays, point to the fact that water may be seen standing directly over the drains, on thorough-drained fields. We have seen this on our own fields. In one instance, we [324] had, after laying tiles through a field, at 50 feet intervals, in the same Autumn, when the land was wet, teamed across it a large quantity of soil for compost, with a heavy ox-team. The next Spring, the water stood for many days in that track which passed across tile-drains, after it had disappeared elsewhere in the field. A fine crop of Indian corn grew on the field that year, but the effect of the puddling was visible the whole season. "One inch of wet and worked clay," says a scientific writer, "will prevent water from passing through, so long as it is kept wet, as effectually as a yard will do."

"If," says Gisborne, "you eat off turnips with sheep, if you plow the land, or cart on it, or in any way puddle it, when it is wet, of course the water will lie on the surface, and will not go to your drains. A four-foot drain may go very near a pit, or a water-course, without attracting water from either, because water-courses almost invariably puddle their beds; and the same effect is produced in pits by the treading of cattle, and even by the motion of the water produced by wind. A very thin film of puddle, always wet on one side, is impervious, *because it cannot crack*."

In those four words, we find an allusion to the whole mystery of the drainage of clays—a key which unlocks the secret by which the toughest of these soils may be converted, as by a fairy charm, to fields of waving grain.

## CRACKING OF CLAYS BY DRYING.

"In drying under the influence of the sun," says Prof. Johnston, "soils shrink in, and thus diminish in bulk, in proportion to the quantity of clay, or of peaty matter, they contain. Sand scarcely diminishes at all in bulk by drying; but peat shrinks one-fifth in bulk, and strong agricultural clay nearly as much." By laying drains in land, we take from it that portion of the water that will run out at the bottom. The sun, by evaporation, then takes out a portion at the top. The soil is thus contracted, and, as the ends of the field cannot approach each other, [325] both soil and subsoil are torn apart, and divided by a network of cracks and fissures. Every one who is familiar with clay land, or who has observed the bottom of a ditch or frog pond by the roadside, must have observed these cracks, thus caused by the contraction of the soil in drying. The same contraction occurs in drier land, by cold, in Winter; by which, in cold regions, deep rents are made in the earth, and reports, like those of cannon, are often heard. The cracking by drying, however, is more quiet in its effects, merely dividing the ground, noiselessly, into smaller and smaller masses, as the process proceeds. Were it not for this process, it may well be doubted whether clay lands could be effectually drained at all. Nature, however, seems to second our efforts here, for we have seen that the stiffer the clay, the greater the contraction, and the more the soil is split up and rendered permeable by this operation.

These cracks are found, by observation, to commence at the drains, and extend further and further, in almost straight lines, into the subsoil, forming so many minor drains, or feeders, all leading to the tiles. These main fissures have numerous smaller ones diverging from them, so that the whole mass is divided and subdivided into the most minute portions. The main fissures gradually enlarge, as the dryness increases, and, at the same time, lengthen out; so that, in a very dry season, they may be traced the whole way between the drains. The following cut will give some idea of these cracks, or fissures, as they exist in a dry time:

[326]

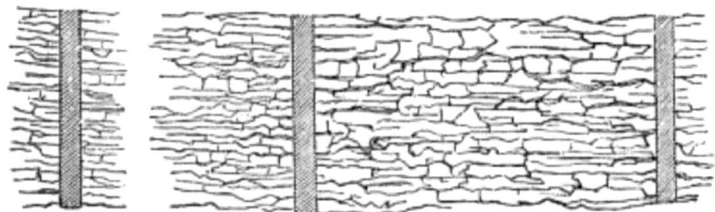

Fig. 98.—Cracking of Clays by Drainage.

Mr. Gisborne says: "Clay lands always shrink and crack with drought; and the stiffer the clay, the greater the shrinking, as brick-makers well know. In the great drought thirty-six years ago, we saw, in a very retentive soil in the Vale of Belvoir, cracks which it was not very pleasant to ride among. This very Summer, on land, which, with reference to this very subject, the owner stated to be impervious, we put a walking-stick three feet into a sun-crack without finding a bottom, and the whole surface was a network of cracks. In the drained soil, the roots follow the threads of vegetable mould which have been washed into the cracks, and get an abiding tenure. Earth-worms follow either the roots or the mould. Permanent schisms are established in the clay, and its whole character is changed."

In the United States, the supply of rain is far less uniform than in England, and much severer droughts are experienced. Thus the contraction, and consequent cracking of the soil, must be greater here than in that country.

In laying drains more than four feet deep, in the stiffest clay which the author has seen, in a neighborhood furnishing abundance of brick and potter's clay, these cracks were seen to extend to the very bottoms of the drains, not in single fissures from top to bottom, but in innumerable seams running in all directions, so that the earth, moved with the pick-axe, came up in little cubes and flakes, and could be separated into pieces of an inch or less diameter. This was on a ridge which received no water except from the clouds, having no springs in or upon it, yet so nearly impervious to water, that it remained soft and muddy till late in June. In Midsummer, however, under our burning sun, it had, by evaporation, been so much dried as to produce the effect described.

In England, we learn, that these cracks extend to the depth of four feet or more. Mr. Hewitt Davis stated in a public discussion, with reference to draining strong soils, that, "he gave four feet as the minimum depth of the drains in these soils, because he had always found that the cracks and fissures formed by the drought and changes of temperature, on the strongest clay, and which made these soils permeable, [327] extended below this depth, and the water from the surface might be made to reach the drains at this distance."

In clay that has never been dried, as for instance, that found under wet meadows from which the water has but recently been drawn, we should not, of course, expect to find these cracks. Accordingly, we find sometimes in clay pits, excavated below the permanent water-line, and in wells, that the clay is in a compact mass, and tears apart without exhibiting anything like these divisions.

We should not expect that, on such a clay, the full effect of drainage would be at once apparent. The water falling on the surface would very slowly find its way downward, at first. But after the heat of Summer, aided by the drains underneath, had contracted and cracked the soil, passages for the water would soon be found, and, after a few years, the whole mass, to the depth of the drains, would become open and permeable. As an old English farmer said of his drains, "They do better year by year; the water gets a habit of coming to them." Although this be not philosophical language, yet the fact is correctly stated. Water tends towards the lowest openings. A deep well often diverts the underground stream from a shallower well, and lays it dry. A single railroad cut sometimes draws off the supply of water from a whole neighborhood. Passages thus formed are enlarged by the pressure of the water, and new ones are opened by the causes already suggested, till the drainage becomes perfect for all practical purposes. So much is this cracking process relied on to facilitate drainage, that skillful drainers frequently leave their ditches partly open, after laying the tiles, that the heat may produce the more effect during the first season.

As to the depth of drains in stiff clays, enough has already been said, under the appropriate title. In England, the weight of authority is in favor of four-foot drains. [328] In this country, a less depth has thus far, in general, been adopted in practice, but it is believed that this has been because a greater depth has not been tried. It is understood, that the most successful drainers in the State of New York, have been satisfied with three-foot drains, not, as it is believed, because there is any instance on record, in this country, of the failure of four-foot drains, but because the effect of more shallow drains has been so satisfactory, that it has been thought a useless expense to go deeper. To Mr. Johnston and to Mr. Delafield, of Seneca County, the country is greatly indebted for their enterprise and leadership in the matter of drainage. Mr. Johnston gives it as his opinion, that "three feet is deep enough, if the bottom is hard enough to lay tiles on; if not, go deeper."

Without intimating that any different mode of drainage than that adopted, would have been better on Mr. Johnston's farm, we should be unwilling to surrender, even to the opinion of Mr. Johnston and his friends, our conviction that, in general, three-foot drains are too shallow. Mr. Johnston expressly disclaims any experience in draining a proper clay soil. In the *Country Gentleman*, of June 10th, 1848, he says:

"In a subsoil that is impervious to water, either by being a red clay, blue clay, or hard-pan, within a foot of the surface, I would recommend farmers to feel their way very cautiously in draining. If tiles and labor were as low here as in Great Britain, we could afford to make drains sixteen feet apart in such land, and then, by loosening the soil, say twenty inches deep, by the subsoil plow, I think such land might be made perfectly dry; but I don't think the time is yet come, considering the cost of tiles and labor, to undertake such an outlay; but still it might pay *in the end*. I have found only a little of red clay subsoil in draining my farm. I never had any blue clay on my farm, or hard-pan, to trouble me; but I can readily perceive that it must be equally bad to drain as the tenacious red clay. If I were going to purchase another farm, I would look a great deal more to the subsoil than the surface soil. If the subsoil is right, the surface soil, I think, cannot be wrong."

[329]

In the same paper, under date of July 8th, Mr. Johnston says, "The only experience I have had in digging into soils, to judge of draining out of this county (Seneca), was in Niagara." He states the result of his observations thus:

"A few inches below the surface I found a stiff blue clay for about ten inches deep, and as impervious to water as so much iron. Underneath that blue clay, I found a red clay, apparently impervious to water; but, as water could not get through the blue, I could only guess at that; and, after spending the greater part of the day, with five men digging holes from four to five feet deep, I found I knew no more how such land could be drained, than a man who had never seen a drain dug. I advised the gentleman to try a few experiments, by digging a few ditches, as I laid them out, and plowing as deep as possible with a subsoil plow, but to get no tile until he saw if he could get a run of water. He paid my traveling expenses, treated me very kindly and I have heard nothing from him since.

"Now, if your correspondent's soil and subsoil is similar to that soil I would advise him to feel his way cautiously in draining. Certainly, no man would be fool enough to dig ditches and lay tile, if there is no water to carry off."

In the *Country Gentleman* of Nov. 18th, 1858, we find an interesting statement, by John S. Pettibone, of Manchester, Vermont, partly in reply to the statement of Mr. Johnston.

The experiment by Mr. Pettibone, showing the increased permeability of clay, merely by the passage of water through it, is very interesting. He says, in his letter to the editor:

"When so experienced a drainer as Mr. Johnston expresses an opinion that some soils cannot be drained, it is important we should know what the soil is which cannot be drained. He uses the word *stiff blue* clay, as descriptive of the soil which cannot be drained. * * *

"I had taken a specimen of what I thought to be *stiff blue clay*. That clay, when wet, as taken out, would hold water about as well as iron: yet, from experiments I have made, I am confident that such clay soil can be drained, and at much less expense than a hard-pan soil. Water will pass through such clay, and the clay become dry; and after [330] it becomes once dry, water will, I am convinced, readily pass down through such stiff blue clay. The specimen was taken about three feet below the surface, and on a level with a brook which runs through this clay soil. I filled a one hundred-pound nail-keg with clay taken from the same place. It was so wet, that by shaking, it came to a level, and water rose to the top of the clay. I had made holes in the bottom of the keg, and set it up on blocks. After twenty-four hours I came almost to the conclusion Mr. Johnston did, that water would not pass through this clay. This trial was during the hot, dry weather last Summer. After some ten or twelve days the clay appeared to be dry. I then made a basin-like excavation in the top of the clay, and put water in, and the water disappeared rather slowly. I filled the basin with water frequently, and the oftener I filled it, the more readily it passed off. I left it for more than a week, when we had a heavy shower. After the shower I examined the keg, and not a drop of water was to be seen. I then took a chisel and cut a hole six inches down. I took out a piece like the one I dried in the house, and laid that up till it was perfectly dry. There was a plain difference between the appearance of the two pieces. The texture, I should say, was quite different. That through which the water had passed, after it had been dried, was more open and porous. It did not possess so much of the blue cast. In less than one hour after the rain fell, the clay taken six inches from the top of the keg would crumble by rubbing in the hand."

When we observe the effect of heat in opening clays to water by cracking, and the effect of the water itself, aided, as it doubtless is, by the action of the air, in rendering the soil permeable, we hardly need feel discouraged if the question rested entirely on this evidence; but when we consider that thousands upon thousands of acres of the stiffest clays have been, in England and Scotland, rescued from utter barrenness by drainage, and made to yield the largest crops, we should regard the question of practicability as settled. The only question left for decision is whether, under all the circumstances of each particular case, the operation of draining our clay lands will be expedient—whether their increased value will pay the expense. It is often objected to deep drains in clays, that it is so far down to the drains that the water cannot readily [331] pass through so large a mass. If we think merely of a drop of rain falling on the surface, and obliged to find its devious way through the mazes of cracks and particles till it gains an outlet at the bottom of four feet of clay, it does seem a discouraging journey for the poor little solitary thing; but there is a more correct view of the matter, which somewhat relieves the difficulty.

All the water that will run out of the soil has departed; but the soil holds a vast amount still, by attraction. The rain begins to fall; and when the soil is saturated, a portion passes into the drain; but it is, by no means, the water which last fell upon the surface, but that which was next the drain before the rain fell. If you pour water into a tube that is nearly full, the water which will first run from the other end is manifestly not that which you pour in. So the ground is full of little tubes, open at both ends, in which the water is held by attraction. A drop upon the surface drives out a drop at the lower end, into to the drain, and so the process goes on—the drains beginning to run as soon as the rain commences, and ceasing to flow only when the principle of attraction balances the power of gravitation.

## PRESSURE OF WATER IN THE SOIL.

In connection with the passage of water through clay soil, it may be appropriate to advert to the question sometimes mooted, whether in a soil filled with water, at four feet depth, there is the same pressure as there would be, at the same depth, in a river or pond. The pressure of fluids on a given area, is, ordinarily, in proportion to their vertical height; and the pressure of a column of water, four feet high, would be sufficient to drive the lower particles into an opening like a drain, with considerable force, and the upper part of such a column would essentially aid the lower part in its downward passage. Does this pressure exist? Mr. Gisborne speaks undoubtingly on this point, thus: [332]

"We will assume the drain to be four feet deep, and the water-table to be at one foot below the surface of the earth. Every particle of water which lies at three feet below the water-table, has on it the pressure of a column of water three feet high. This pressure will drive the particle in any direction in which it finds no resistance, with a rapidity varying inversely to the friction of the medium through which the column acts. The bottom of our drains will offer no resistance, and into it particles of water will be pushed, in conformity with the rule we have stated; rapidly, if the medium opposes little friction; slowly, if it opposes much. The water so pushed in runs off by the drain, the column of pressure being diminished in proportion to the water which runs off."

Mr. Thomas Arkell, in a paper read before the Society of Arts, in 1855, says, on this point:

"The pressure due to a head of water of four or five feet, may be imagined from the force with which water will come through the crevices of a hatch, with that depth of water above it. Now, there is the same pressure of water to enter the vacuum in the pipe-drain, as there is against the hatches, supposing the land to be full to the surface."

We do not find any intimation that there is any error in the view advanced by the learned gentleman quoted; and if there is none, we have an explanation of the faculty which water seems to have, of finding its way into drainpipes. Yet, we feel bound to confess, that, aside from authority, we should have supposed that the pressure due to a column of pure water, would be essentially lessened, by the interposition of solid matter between its particles.

# CHAPTER XX [333]
# EFFECT OF DRAINAGE ON STREAMS AND RIVERS.

Drainage Hastens the Supply to the Streams, and thus Creates Freshets. — Effect of Drainage on Meadows below; on Water Privileges. — Conflict of Manufacturing and Agricultural Interests. — English Opinions and Facts. — Uses of Drainage Water. — Irrigation. — Drainage Water for Stock. — How used by Mr. Mechi.

The effect of drainage upon streams and rivers, has, perhaps, little to interest merely practical men, in this country, at present; but the time will soon arrive, when mill-owners and land-owners will be compelled to investigate the subject. Men unaccustomed to minute investigation, are slow to appreciate the great effects produced by apparently small causes; and it may seem to many, that the operations of drainage for agriculture, are too insignificant in their details, perceptibly to affect the flow of mill-streams and rivers. A moment's thought will convince the most skeptical, that the thorough-drainage of the wet lands, even of a New England township, must produce sensible effects upon the streams which convey its surplus water toward the sea.

In making investigations to ascertain what quantity of water may be relied upon to supply a reservoir, whether natural or artificial, for the use of a town or city, a survey is first taken of the district of territory which naturally is drained into the reservoir, and thus the number of square miles of surface is ascertained. Then the rain-tables are consulted, and the fall of rain upon the surveyed district [334] is computed. The ascertained proportion of rain-fall, which usually goes off by evaporation, is then deducted, which leaves with sufficient accuracy, the amount of water which flows both upon the surface, and through the soil, to the reservoir. With proper deductions for waste by freshets, when the water will overflow the reservoir, and for other known losses, a reliable estimate is readily made, in advance, of the quantity of water supplied to the reservoir.

Now, these reservoirs Nature has placed in all our valleys, in the form of lakes and ponds, and the drainage into them is by natural

357

springs and streams; and the annual amount of the water thus naturally flowing into them may be readily computed, if the area within their head-waters be known. If the earth's surface were, like iron, impervious to water, the rain-water would come in torrents down the hill-sides, and along the gentle declivities, into the streams, creating freshets and inundations in a few hours. But instead of that, the soft showers fall, often on the open, thirsty soil, and so are gradually absorbed. A part of the rain-water is there held, until it returns by evaporation, to the clouds, while a part slowly percolates downward, finding its way into swamps and springy plains, and finally, after days or weeks of wandering, slowly, but surely, finds its outlet in the stream or pond.

If now, this surplus of water, this part which cannot be evaporated, and must therefore, sooner or later, enter the stream or pond, be, by artificial channels, carried directly to its destination, without the delay of filtration through swamps and clay-banks; the effect of rain to raise the streams and ponds, must be more sudden and immediate. Agricultural drains furnish those artificial channels. The flat and mossy swamp, which before retained the water until the Midsummer drought, and then slowly parted [335] with it, by evaporation or gradual filtration, now, by thorough-drainage, in two or three days at most, sends all its surplus water onward to the natural stream. The stagnant clay-beds, which formerly, by slow degrees, allowed the water to filter through them to the wayside ditch, and then to the river, now, by drainage, contribute their proportion, in a few hours, to swell the stream. Thus, evaporation is lessened, and the amount of water which enters the natural channels largely increased; and, what is of more importance, the water which flows from the land is sent at once, after its fall from the heavens, into the streams. This produces upon the mill-streams a two-fold effect; first, to raise sudden freshets to overflow the dams, and sweep away the mills; and, secondly, to dry up their supply in dry seasons, and to diminish their water-power.

Upon the low meadows which border the streams, the effects of the drainage of lands above them are various, according to their position. In many cases, it must subject them to inundation by Summer freshets, and must require for their protection, catch-waters and embankments, and large facilities for drainage.

The effect of drainage upon "water privileges," must inevitably be, to lessen their value, by giving them a sudden surplus, followed by drought, instead of a regular supply of water. Water-power companies and mill-owners are never careless of their interests. Through the patriotic desire to foster home-manufactures, our State legislatures have granted many peculiar privileges to manufacturing corporations. Indeed, all the streams and rivers of New England are chained to labor at their wheels.

Agriculture has thus far taken care of herself, but is destined soon to come in collision with the chartered privileges of manufactures. Many questions, touching the right of land-owners to change the natural flow of the [336] water, to the injury of mill-owners; many questions touching the right of mill-owners to obstruct the natural course of streams, to the injury of the farmer, will inevitably arise in our Courts. Slowly, and step by step, must the lesser interest of manufactures, recede before the advance of the great fundamental interest of agriculture, until, in process of time, steam, or some yet undiscovered giant power, shall put its hand to the great wheel of the factory and the mill, and the pent-up waters shall subside to their natural banks.

That these are not mere speculations of our own, may be seen from extracts which will be given from answers returned by distinguished observers of these matters in England and Scotland, to a question proposed to them as to the actual effects produced by extensive drainage. Some diversity of opinion is observable in the different replies, which were made, independently in writing, and so are more valuable.

*Mr. Smith.* — "During dry periods, more particularly in Summer, the water in the streams is greatly lessened by thorough-draining; for there is so great a mass of comparatively dry and absorbent soil to receive the rain, that Summer showers, unless very heavy and continuous, will be entirely absorbed."

*Mr. Parkes.* — "The intention and effect of a complete and systematic under-drainage is the liberation of the water of rain more quickly from the land than if it were not drained; and therefore the natural vents, or rivers, very generally require enlargement or deepening, in order to pass off the drainage water in sufficiently quick time, and so as to avoid flooding lower lands.

"The sluggish rivers of the midland and southern counties of England especially, oppose great obstacles to land-drainage, being usually full to the banks, or nearly so, and converted into a series of ponds, by mill-dams erected at a few miles distance below each other; so that, frequently, no effectual drainage of the richest alluvial soil composing the meadows, can be made, without forming embankments, or by pumping, or by resort to other artificial and expensive means.

"The greater number of the corn and other water-mills throughout England ought to be demolished, for the advantage of agriculture, and [337] steam-power should to be provided for the millers. I believe that such an arrangement would, in most cases, prove to be economical both to the landholder and the miller.

"Every old authority, and all modern writers on land drainage in England, have condemned water-mills and mill-dams: and if all the rivers of England were surveyed from the sea to their source, the mills upon them valued, the extent of land injured or benefitted by such mill-dams ascertained, and the whole question of advantage or injury done to the land-owner appreciated and appraised, I have little doubt but that the injury done, would be found so greatly to exceed the rental of the mills, deduction being made of the cost of maintaining them, that it would be a measure of national economy, to buy up the mills, and give the millers steam-power."

*Mr. Spooner.* — "The effect which extensive drainage produces on the main water-courses of districts, is that of increasing the height of their rise at flood times, and rendering the flow and subsidence more rapid than before. I have repeatedly heard the River Tweed adduced as a striking instance of this fact, and that the change has taken place within the observation of the present generation."

*Mr. Maccaw.* — "It has been observed that, after extensive surface-drainage on the sheepwalks in the higher parts of the country, and when the lower lands were enclosed by ditches, and partially drained for the purposes of cultivation, all rivers flowing therefrom, rise more rapidly after heavy rains or falls of snow, and discharge their surplus waters more quickly, than under former circumstances."

*Mr. Beattie.* — "It renders them more speedily flooded, and to a greater height, and they fall sooner. Rivers are lower in Summer and higher in Winter."

*Mr. Nielson.* — "The immediate effect of the drainage of higher lands has often been to inundate the lower levels."

In a prize essay of John Algernon Clarke, speaking of the effect of drainage along the course of the River Nene, in England, he says:

"The upland farms are delivering their drain-water in much larger quantities, and more immediately after the downfall, than formerly, and swelling to the depth of three to six feet over the 20,000 acres of open ground, which form one vast reservoir for it above and below Peterborough. The Nene used to overflow its banks, to the extreme height, about the third day after rain: the floods now reach the same height in about half that time. Twelve hours' rain will generally cause an overflow of the land, which all lies unembanked from the stream; and where it [338] is already saturated, this takes place in six or even in two hours. Such a quick rise will cause one body of flood-water to extend for forty or fifty miles in succession, with a width varying from a quarter of a mile to a mile; but it stays sometimes for six weeks, or even two months, upon the ground. And those floods come down with an alarming power and velocity — bridges which have stood for a century are washed away, and districts where floods were previously unknown have became liable to their sudden periodical inundations. The land being wholly in meadow, suffers very heavily from the destruction of its hay. So sudden are the inundations, that it frequently happens that hay made in the day has, in the night been found swimming and gone. A public-house sign at Wansford commemorates the locally-famed circumstance of a man who, having fallen asleep on a hay-cock, was carried down the stream by a sudden flood: awakening just under the bridge of that town, and being informed where he was, he demanded, in astonishment, if this were 'Wansford in England.'"

The fact that the floods in that neighborhood now reach their height in half their former time, in consequence of the drainage of the "upland farms," is very significant.

Mr. Denton thus speaks upon the same point, though his immediate subject was that of compulsory outfalls.

"Although the quantity of land drained was small, in comparison to that which remained to be drained, the water which was discharged by the drainage already effected found its way so rapidly to the outfalls, that the consequences were becoming more and more injurious every day. The millers were now suffering from two causes. At times of excess, after a considerable fall of rain, and when the miller was injuriously overloaded, the excess was increased by the rapidity with which the under-drains discharged themselves; and as the quantity of water thus discharged, must necessarily lessen the subsequent supply, the period of drought was advanced in a corresponding degree. As the millers already saw this, and were anticipating increasing losses, they would join in finding a substitute for water-power upon fair terms."

It is not supposed, that any considerable practical effects of drainage, upon the streams of this country, have been observed. A treatise, however, upon the general subject of Drainage, which should omit a point like this, which must, before many years, attract serious attention, would be quite incomplete. Whether the effect of a system of [339] thorough-drainage make for or against the interest of mill and meadow owners on the lower parts of streams, should have no influence over those who design only to present the truth, in all its varied aspects.

As some compensation for the evils which may fall upon lands at a lower level, by drainage of uplands, it may be interesting to notice briefly in this place, some of the uses to which drainage-water has been applied, for the advantage of lower lands. In many cases, in Great Britain, the water of drainage has been preserved in reservoirs, or artificial ponds, and applied for the irrigation of water meadows; and as is suggested by Lieut. Maury, in a letter quoted in our introductory chapter, the same may, in many localities, be done in this country, and thus our crops of grass be often tripled, on our low meadows. In many cases, water from deep drains, will furnish the most convenient supply for barn yards and pastures. It is usually sufficiently pure and cool in Summer, and is preferred by cattle to the water of running streams.

On Mr. Mechi's farm at Tiptree Hall, in England, we observed a large cistern, in which all the manure necessary for the highest culture of 170 acres of land, is liquified, and from which it is pumped out by a steam engine, over the farm. All the water, which supplies the cistern, is collected from tile drains on the farm, where there had before been no running water.

# CHAPTER XXI [340]
# LEGISLATION—DRAINAGE COMPANIES.

England protects her Farmers.—Meadows ruined by Corporation dams.—Old Mills often Nuisances.—Factory Reservoirs.—Flowage extends above level of Dam.—Rye and Derwent Drainage.—Give Steam for Water-Power.—Right to Drain through land of others.—Right to natural flow of Water.—Laws of Mass.—Right to Flow; why not to Drain?—Land-drainage Companies in England.—Lincolnshire Fens.—Government Loans for Drainage.

Nothing more clearly shows the universal interest and confidence of the people of Great Britain, in the operation of land-drainage, than the acts of Parliament in relation to the subject. The conservatism of England, in the view of an American, is striking. She never takes a step till she is sure she is right. Justly proud of her position among the nations, she deems change an unsafe experiment, and what has been, much safer than what might be. Vested rights are sacred in England, and especially rights in lands, which are emphatically real estate there.

Such are the sentiments of the people, and such the sentiments of their representatives and exponents, the Lords and Commons.

Yet England has been so impressed with the importance of improving the condition of the people, of increasing the wealth of the nation, of enriching both tenant and landlord, by draining the land, that the history of her legislation, in aid of such operations, affords a lesson of progress even to fast Young America. Powers have been granted, by which encumbered estates may be charged with the expenses of drainage, so that remainder-men and reversioners, without their consent, shall be compelled to contribute to present improvements; so that careless or obstinate adjacent proprietors shall be compelled to keep open their ditches for outfalls to their neighbor's drains; so that mill-dams, and other obstructions to the natural flow of the water, may be removed for the benefit of agriculture; and, finally, the Government has itself [341] furnished funds, by way of loans, of millions of pounds, in aid of improvements of this character.

In America, where private individual right is usually compelled to yield to the good of the whole, and where selfishness and obstinacy do not long stand in the pathway of progress, obstructing manifest improvement in the condition of the people; we are yet far behind England in legal facilities for promoting the improvement of land culture. This is because the attention of the public has not been particularly called to the subject.

Manufacturing corporations are created by special acts of legislation. In many States, rights to flow, and ruin, by inundation, most valuable lands along the course of rivers, and by the banks of ponds and lakes, to aid the water-power of mills, are granted to companies, and the land-owner is compelled to part with his meadows for such compensation as a committee or jury shall assess.

In almost every town in New England there are hundreds, and often thousands, of acres of lands, that might be most productive to the farmer; overflowed half the year with water, to drive some old saw-mill, or grist-mill, or cotton-mill, which has not made a dividend, or paid expenses, for a quarter of a century. The whole water-power, which, perhaps, ruins for cultivation a thousand acres of fertile land, and divides and breaks up farms, by creating little creeks and swamps throughout all the neighboring valleys, is not worth, and would not be assessed, by impartial men, at one thousand dollars. Yet, though there is power to take the farmer's land for the benefit of manufacturers, there is no power to take down the company's dam for the benefit of agriculture. An old saw-mill, which can only run a few days in a Spring freshet, often swamps a half-township of land, because somebody's great-grandfather had a prescriptive right to flow, when lands were of no value, and saw-mills were a public blessing.

There are numerous cases, within our own knowledge, where the very land overflowed and ruined by some incorporated company, would, if allowed to produce its natural growth of timber and wood, furnish ten times the fuel necessary to supply steam-engines, to propel the machinery carried by the water-power.

Not satisfied with obstructing the streams in their course, the larger companies are, of late, making use of the interior lakes, fifty or a hundred miles inland, as reservoirs, to keep back water for the use of the mills in the summer droughts. Thus are thousands of acres of land drowned, and rendered worse than useless; for the water is kept up till Midsummer, and drawn off when a dog-day climate is just ready to [342] convert the rich and slimy sediment of the pond into pestilential vapors. These waters, too, controlled by the mill-owners, are thus let down in floods, in Midsummer, to overflow the meadows and corn-fields of the farmer, or the intervals and bottom-lands below.

Now, while we would never advocate any attack upon the rights of mill-owners, or ask them to sacrifice their interests to those of agriculture, it surely is proper to call attention to the injury which the productive capacity of the soil is suffering, by the flooding of our best tracts, in sections of country where land is most valuable. Could not mill-owners, in many instances, adopt steam instead of water-power, and becoming land-*draining* companies, instead of land-*drowning* companies; at least, let Nature have free course with her gently-flowing rivers, and allow the promise to be fulfilled, that the earth shall be no more cursed with a flood.

We would ask for the land-owner, simply equality of rights with the mill-owner. If a legislature may grant the right to flow lands, against the will of the owner, to promote manufactures, the same legislature may surely grant the right, upon proper occasion, to remove dams, and other obstructions to our streams, to promote agriculture. The rights of mill-owners are no more sacred than those of land-owners; and the interests of manufactures are, surely, no more important than those of agriculture.

We would not advocate much interference with private rights. In some of the States, no special privileges have been conferred upon water-power companies. They have been left to procure their rights of flowage, by private contract with the land-owners; and in such States, probably, the legislatures would be as slow to interfere with rights of flowage, as with other rights. Yet, there are cases where, for the preservation of the health of the community, and for the general convenience, governments have everywhere exercised the power of interfering with private property, and limiting the control of the owners. To preserve the public health, we abate as nuisances, by process of law, slaughter-houses, and other establishments offensive to health and comfort, and we provide, by compulsory assessments upon land-owners, for sewerage, for side-walks, and the like, in our cities.

Everywhere, for the public good, we take private property for highways, upon just compensation, and the property of corporations is thus taken, like that of individuals.

Again, we compel adjacent owners to fence their lands, and maintain their proportion of division fences of the legal height, and we elect fence viewers, with power to adjust equitably, the expenses of such [343] fences. We assess bachelors and maidens, in most States, for the construction of schoolhouses, and the education of the children of others, and, in various ways, compel each member of society to contribute to the common welfare.

How far it may be competent, for a State legislature to provide for, or assist in, the drainage of extensive and unhealthy marshes; or how far individual owners should be compelled to contribute to a common improvement of their lands; or how far, and in what cases, one land-owner should be authorized to enter upon land of another, to secure or maintain the best use of his own land—these are questions which it is unnecessary for us to attempt to determine. It is well that they should be suggested, because they will, at no distant day, engage much attention. It is well, too, that the steps which conservative England has thought it proper to take in this direction, should be understood, that we may the better determine whether any, and if any, what course our States may safely take, to aid the great and leading interest of our country.

The swamps and stagnant meadows along our small streams and our rivers, which are taken from the farmer, by flowage, for the benefit of mills, are often, in New England, the most fertile part of the townships—equal to the bottom lands of the West; and they are right by the doors of young men, who leave their homes with regret, because the rich land of far-off new States offers temptations, which their native soil cannot present.

It is certainly of great importance to the old States, to inquire into these matters, and set proper bounds to the use of streams for water-powers. The associated wealth and influence of manufacturers, is always more powerful than the individual efforts of the land-owners.

Reservoirs are always growing larger, and dams continually grow higher and tighter. The water, by little and little, creeps insidiously on to, and into, the meadows far above the obstruction, and the land-owner must often elect between submission to this aggression, and a tedious law-suit with a powerful adversary. The evil of obstructions to streams and rivers, is by no means limited to the land visibly flowed, nor to land at the level of the dam. Running water is never level, or it could not flow; and in crooked streams, which flow through meadows, obstructed by grass and bushes, the water raised by a dam, often stands many feet higher, at a mile or two back, than at the dam. It is extremely difficult to set limits to the effect of such a flowage. Water is flowed into the subsoil, or rather is prevented from running out; the natural drainage of the country is prevented; and land which might well [344] be drained artificially, were the stream not obstructed, is found to lie so near the level, as to be deprived of the requisite fall by back water, or the sluggish current occasioned by the dam.

These obstructions to drainage have become subjects of much attention, and of legislative intervention in various forms in England, and some of the facts elicited in their investigations are very instructive.

In a discussion before the Society of Arts, in 1855, in which many gentlemen, experienced in drainage, took a part, this subject of obstruction by mill-dams came up.

Mr. G. Donaldson said he had been much engaged in works of land-drainage, and that, in many instances, great difficulties were experienced in obtaining outfalls, owing to the water rights, on the course of rivers for mill-power, &c.

Mr. R. Grantham spoke of the necessity of further legislation, "so as to give power to lower bridges and culverts, under public roads, and straighten and deepen rivers and streams." But, he said, authority was wanting, above all, "for the removal of mills, dams, and other obstructions in rivers, which, in many cases, did incalculable injury, many times exceeding the value of the mills, by keeping up the level of rivers, and rendering it totally impossible to drain the adjoining lands."

Mr. R. F. Davis said, "If they were to go into the midland districts, they would see great injury done, from damming the water for mills."

In Scotland, the same difficulty has arisen. "In many parts of this country," says a Scottish writer, "small lochs (lakes) and dams are kept up, for the sake of mills, under old tenures, which, if drained, the land gained by that operation, would, in many instances, be worth ten times the rent of such mills."

In the case of the Rye and Derwent Drainage, an account of which is found in the 14th Vol. of the Journal of the Royal Agricultural Society, a plan of compensation was adopted, where it became necessary to remove dams and other obstructions, which is worthy of attention. The Commissioners under the Act of 1846, removed the mill-wheels, and substituted steam-engines corresponding to the power actually used by the mills, compensating, also, the proprietors for inconvenience, and the future additional expensiveness of the new power.

"The claims of a short canal navigation, two fisheries, and tenants' damages through derangement of business during the alterations, were disposed of without much outlay; and the pecuniary advantages of the work are apparent from the fact, that a single flood, such as frequently [345] overflowed the land, has been known to do more damage, if fairly valued in money, than the whole sum expended under the act."

Under this act, it became necessary for the Commissioners to estimate the comparative cost of steam and water-power, in order to carry out their idea of giving to the mill-owners a steam-power equivalent to their water-power.

"As the greater part of their water-power was employed on corn and flour-mills, upon these the calculations were chiefly based. It was generally admitted to be very near the truth, that to turn a pair of flour-mill stones properly, requires a power equal to that of two-and-a-half horses, or on an average, twenty horses' power, to turn and work a mill of eight-pairs of stones, and that the total cost of a twenty-horse steam-engine, with all its appliances, would be $5,000, or $250 per horse power."

Calculations for the maintenance of the steam-power are also given; but this depends so much on local circumstances, that English estimates would be of little value to us.

The arrangements in this case with the mill-owners, were made by contract, and not by force of any arbitrary power, and the success of the enterprise, in the drainage of the lands, the prevention of damage by floods, especially in hay and harvest-time, and in the improvement of the health of vegetation, as well as of man and animals, is said to be strikingly manifest.

This act provides for a "water-bailiff," whose duty it is to inspect the rivers, streams, water-courses, &c., and enforce the due maintenance of the banks, and the uninterrupted discharge of the waters at all times.

## COMPULSORY OUTFALLS.

It often happens, especially in New England, where farms are small, and the country is broken, that an owner of valuable lands overcharged with water, perhaps a swamp or low meadow, or perhaps a field of upland, lying nearly level, desires to drain his tract, but cannot find sufficient fall, without going upon the land of owners below. These adjacent owners may not appreciate the advantages of drainage; or their lands may not require it; or, what is not unusual, they may from various motives, good and evil, refuse to allow their lands to be meddled with.

Now, without desiring to be understood as speaking judicially, we know of no authority of law by which a land-owner may enter upon the territory of his neighbor for the purpose of draining his own land, and perhaps no such power should ever be conferred. All owners upon streams, great and small, have however, the right to the natural flow [346] of the water, both above and below. Their neighbors below cannot obstruct a stream so as to flow back the water upon, or into, the land above; and where artificial water-courses, as ditches and drains have long been opened, the presumption would be that all persons benefitted by them, have the right to have them kept open.

Parliament is held to be omnipotent, and in the act of 1847, known as Lord Lincoln's Act, its power is well illustrated, as is also the determination of the British nation that no trifling impediments shall hinder the progress of the great work of draining lands for agriculture. The act, in effect, authorizes any person interested in draining his lands, to clear a passage through all obstructions, wherever it would be worth the expense of works and compensation.

Its general provisions may be found in the 15th Vol. of the Journal of the Royal Agricultural Society.

It is not the province of the author, to decide what may properly be done within the authority of different States, in aid of public or private drainage enterprises. The State Legislatures are not, like Parliament, omnipotent. They are limited by their written constitutions. Perhaps no better criterion of power, with respect to compelling contribution, by persons benefitted, to the cost of drainage, and with interfering with individual rights, for public or private advantage, can be found, than the exercise of power in the cases of fences and of flowage.

If we may lawfully compel a person to fence his land, to exclude the cattle of other persons, or, if he neglect to fence, subject him to their depredations, without indemnity, as is done in many States; or if we may compel him to contribute to the erection of division fences, of a given height, though he has no animal in the world to be shut in or out of his field, there would seem to be equal reason, in compelling him to dig half a division ditch for the benefit of himself and neighbor.

If, again, as we have already hinted, the Legislature may authorize a corporation to flow and inundate the land of an unwilling citizen, to raise a water-power for a cotton-mill, it must be a nice discrimination of powers, that prohibits the same Legislature from authorizing the entry into lands of a protesting mill-owner, or of an unknown or cross-grained proprietor, to open an outlet for a valuable, health-giving system of drainage.

In the valuable treatise of Dr. Warder, of Cincinnati, recently published in New York, upon Hedges and Evergreens, an abstract is given of the statutes of most of our States, upon the subject of fences, and we know of no other book, in which so good an idea of the legislation on this subject, can be so readily obtained. [347]

By the statutes of Massachusetts, any person may erect and maintain a water-mill, and dam to raise water for working it, upon and across any stream that is not navigable, provided he does not interfere with existing mills. Any person whose land is overflowed, may, on complaint, have a trial and a verdict of a jury; which may fix the height of the dam, decide whether it shall be left open any part of the year, and fix compensation, either annual or in gross, for the injury. All other remedies for such flowage are taken away, and thus the land of the owner may be converted into a mill-pond against his consent.

We find nothing in the Massachusetts statutes which gives to land-owners, desirous of improving their wet lands, any power to interfere in any way with the rights of mill-owners, for the drainage of lands. The statutes of the Commonwealth, however, make liberal and stringent provisions, for compelling unwilling owners to contribute to the drainage of wet lands.

For the convenience of those who may be desirous of procuring legislation on this subject, we will give a brief abstract of the leading statute of Massachusetts regulating this matter. It may be found in Chapter 115 of the Revised Statutes, of 1836. The first Section explains the general object.

When any meadow, swamp, marsh, beach, or other low land, shall be held by several proprietors, and it shall be necessary or useful to drain or flow the same, or to remove obstructions in rivers or streams leading therefrom, such improvements may be effected, under the direction of Commissioners, in the manner provided in this chapter.

The statute provides that the proprietors, or a greater part of them in interest, may apply, by petition, to the Court of Common Pleas, setting forth the proposed improvements, and for notice to the proprietors who do not join in the petition, and for a hearing. The court may then appoint three, five, or seven commissioners to cause the improvements to be effected. The commissioners are authorized to cause dams or dikes to be erected on the premises, at such places, and in such manner as they shall direct; and may order the land to be flowed thereby, for such periods of each year as they shall think most beneficial, and also cause ditches to be opened on the premises, and obstructions in any rivers or streams leading therefrom to be removed.

Provision is made for assessment of the expenses of the improvements, upon all the proprietors, according to the benefit each will derive from it, and for the collection of the amount assessed.

"When the commissioners shall find it necessary or expedient to reduce or raise the waters, for the purpose of obtaining a view of the premises, or for the more convenient or expeditious removal of obstructions [348] therein, they may open the flood-gates of any mill, or make other needful passages through or round the dam thereof or erect a temporary dam on the land of any person, who is not a party to the proceedings, and may maintain such dam, or such passages for the water, as long as shall be necessary for the purposes aforesaid."

Provision is made for previous notice to persons who are not parties, and for compensation to them for injuries occasioned by the interference, and for appeal to the courts.

This statute gives, by no means, the powers necessary to compel contribution to all necessary drainage, because, first, it is limited in its application to "meadow, swamp, marsh, beach, or other low land." The word *meadow*, in New England, is used in its original sense of flat and wet land. Secondly, the statute seems to give no authority to open permanent ditches on the land of others than the owners of such low land, although it provides for temporary passages for the purposes of "obtaining a view of the premises, or for the more convenient or expeditious removal of obstructions *therein*" — the word "therein" referring to the "premises" under improvement, so that there is no provision for outfalls, under this statute, except through natural streams.

By a statute of March 28, 1855, the Legislature of Massachusetts has exercised a *power* as extensive as is desirable for all purposes of drainage, although the provisions of the act referred to are not, perhaps, so broad as may be found necessary, in order to open outfalls and remove all obstructions to drainage. As this act is believed to be peculiar, we give its substance:

"An Act to authorize the making of Roads and Drains in certain cases.

"Sect. 1. Any town or city, person or persons, company or body corporate, having the ownership of low lands, lakes, swamps, quarries, mines, or mineral deposits, that, by means of adjacent lands belonging to other persons, or occupied as a highway, cannot be approached, worked, drained, or used in the ordinary manner without crossing said lands or highway, may be authorized to establish roads, drains, ditches, tunnels, and railways to said places in the manner herein provided.

"Sect. 2. The party desiring to make such improvements shall file a petition therefor with the commissioners of the county in which the premises are situated, setting forth the names of the persons interested, if known to the petitioner, and also, in detail, the nature of the proposed improvement, and the situation of the adjoining lands."

Sect. 3 provides for notice to owners and town authorities.

Sect. 4 provides for a hearing, and laying out the improvement, and [349] assessment of damages upon the respective parties, "having strict regard to the benefits which they will receive."

Sect. 5 provides for repairs by a majority of those benefitted; and Sect. 6 for appeals, as in the case of highways.

By an act of 1857, this act was so far amended as to authorize the application for the desired improvement, to be made to the Selectmen of the town, or the Mayor and Aldermen of the city, in case the lands over which the improvement is desired are all situated in one town or city.

It is manifest certainly, that the State assumes power sufficient to authorize any interference with private property that may be necessary for the most extended and thorough drainage operations. The power which may compel a man to improve his portion of a swamp, may apply as well to his wet hill-sides; and the power which may open temporary passages through lands or dams, without consent of the owner, may keep them open permanently, if expedient.

375

# LAND DRAINAGE COMPANIES.

Besides the charters which have at various times, for many centuries, been granted to companies, for the drainage of fens and marshes, and other lowlands, in modern times, great encouragement has been given by the British Government for the drainage and other improvement of high-lands. Not only have extensive powers been granted to companies, to proceed with their own means, to effect the objects in view, but the Government itself has advanced money, by way of loan, in aid of drainage and like improvements.

By the provisions of two acts of Parliament, no less than $20,000,000 have been loaned in aid of such improvements. These acts are generally known as Public Moneys Drainage Acts. There are already four chartered companies for the same general objects, doing an immense amount of business, on *private* funds.

It will be sufficient, perhaps, to state, in general terms, the mode of operation under these several acts.

Most lands in England are held under incumbrances of some kind. Many are entailed, as it is termed: that is to say, vested for life in certain persons, and then to go to others, the tenant for life having no power to sell the property. Often, the life estate is owned by one person, and the remainder by a stranger, or remote branch of the family, whom the life-tenant has no desire to benefit. In such cases, the tenant, or occupant, would be unwilling to make expensive improvements at his own cost, which might benefit himself but a few years, and then go into other hands. [350]

On the other hand, the remainder-man would have no right to meddle with the property while the tenant-for-life was in possession; and it would be rare, that all those interested could agree to unite in efforts to increase the general value of the estate, by such improvements.

The great object in view was, then, to devise means, by which such estates, suffering for want of systematic, and often expensive, drainage operations, might be improved, and the cost of improvement be charged on the estate, so as to do no injustice to any party interested.

The plan finally adopted, is, to allow the tenant or occupant to have the improvement made, either by expending his private funds, or by borrowing of the Government or the private companies, and having the amount expended, made a charge on the land, to be paid, in annual payments, by the person who shall be in occupation each year. Under one of these acts, the term of payment is fixed at 22 years, and under a later act, at 50 years.

Thus, if A own a life-estate in lands, and B the remainder, and the estate needs draining, A may take such steps as to have the improvement made, by borrowing the money, and repaying it by yearly payments, in such sums as will pay the whole expenditure, with interest, in twenty-two or fifty years: and if A die before the expiration of the term, the succeeding occupants continue the payments until the whole is paid.

A borrows, for instance, $1,000, and expends it in draining the lands. It is made a charge, like a mortgage, on the land, to be paid in equal annual payments for fifty years. At six per cent., the annual payment will be but about $63.33, to pay the whole amount of debt and the interest, in fifty years. A pays this sum annually as long as he lives, and B then takes possession, and pays the annual installment.

If the tenant expend his own money, and die before the whole term expire, he may leave the unpaid balance as a legacy, or part of his own estate, to his heirs.

The whole proceeding is based upon the idea, that the rent or income of the property is sufficiently increased, to make the operation advantageous to all parties. It is assumed, that the operation of drainage, under one of these statutes, will be effectual to increase the rent of the land, to the amount of this annual payment, for at least fifty years. The fact, that the British Government, after the most thorough investigation, has thus pronounced the opinion, that drainage works, properly conducted, will thus increase the rent of land, and remain in full operation a half century at least, affords the best evidence possible, both of the utility and the durability of tile drainage.

# CHAPTER XXII [351]
# DRAINAGE OF CELLARS.

Wet Cellars Unhealthful. — Importance of Cellars in New England. — A Glance at the Garret, by way of Contrast. — Necessity of Drains. — Sketch of an Inundated Cellar. — Tiles best for Drains. — Best Plan of Cellar Drain; Illustration. — Cementing will not do. — Drainage of Barn Cellars. — Uses of them. — Actual Drainage of a very Bad Cellar described. — Drains Outside and Inside; Illustration.

No person needs to be informed that it is unhealthful, as well as inconvenient, to have water, at any time of the year, in the cellar. In New England, the cellar is an essential part of the house. All sorts of vegetables, roots, and fruit, that can be injured by frost, are stored in cellars; and milk, and wine, and cider, and a thousand "vessels of honor," like tubs and buckets, churns and washing-machines, that are liable to injury from heat or cold, or other vicissitude of climate, find a safe retreat in the cellar. Excepting the garret, which is, as Ariosto represents the moon to be, the receptacle of all things useless on earth, the cellar is the greatest "curiosity shop" of the establishment.

The poet finds in the moon,

> "Whate'er was wasted in our earthly state,
> Here safely treasured — each neglected good,
> Time squandered, and occasion ill-bestowed;
> There sparkling chains he found, and knots of gold,
> The specious ties that ill-paired lovers hold;
> Each toil, each loss, each chance that men sustain,
> Save Folly, which alone pervades them all,
> For Folly never quits this earthly ball."
> [352]

In the garret, are the old spinning wheel, the clock reel, the linen wheel with its distaff, your grandfather's knapsack and cartridge-box and Continental coat, your great-aunt's Leghorn bonnet and side-saddle, or pillion, great files of the village newspapers — the

"*Morning Cry*" and "*Midnight Yell*," besides worn out trunks and boxes without number. In the cellar, are the substantiate—barrels of beef, and pork, and apples, "taters" and turnips; in short, the Winter stores of the family.

Many, perhaps most, of the cellars in New England are in some way drained, usually by a stone culvert, laid a little lower than the bottom of the cellar, into which the water is conducted, in the Spring, when it bursts through the walls, or rises at the bottom, by means of little ditches scooped out in the surface.

In some districts, people seem to have little idea of drains, even for cellars; and on flat land, endeavor to set their houses high enough to have their cellars above ground. This, besides being extremely inconvenient for passage out of, and into the house, often fails to make a dry cellar, for the water from the roof runs in, and causes a flood. And such accidents, as they are mildly termed by the improvident builders, often occur by the failure of drains imperfectly laid.

No child, who ever saw a cellar afloat, during one of these inundations, will ever outgrow the impression. You stand on the cellar stairs, and below is a dark waste of waters, of illimitable extent. By the dim glimmer of the dip-candle, a scene is presented which furnishes a tolerable picture of "chaos and old night," but defies all description. Empty dry casks, with cider barrels, wash-tubs, and boxes, ride triumphantly on the surface, while half filled vinegar and molasses kegs, like water-logged ships, roll heavily below. Broken boards and planks, old hoops, and staves, and barrel-heads innumerable, are buoyant [353] with this change of the elements; while floating turnips and apples, with, here and there, a brilliant cabbage head, gleam in this subterranean firmament, like twinkling stars, dimmed by the effulgence of the moon at her full. Magnificent among the lesser vessels of the fleet, "like some tall admiral," rides the enormous "mash-tub," while the astonished rats and mice are splashing about at its base in the dark waters, like sailors just washed, at midnight, from the deck, by a heavy sea.

The lookers-on are filled with various emotions. The farmer sees his thousand bushels of potatoes submerged, and devoted to speedy decay; the good wife mourns for her diluted pickles, and apple sauce, and her drowned firkins of butter; while the boys are anxious to embark on a raft or in the tubs, on an excursion of pleasure and discovery.

To avoid such scenes as the above, every cellar which is not upon a dry sandbank, should be provided with a drain of some kind, which will be at all times, secure.

For a main drain from the cellar, four or six-inch tiles are abundantly sufficient, and where they can be reasonably obtained, much cheaper than stone. The expense of excavation, of hauling stone, and of laying them, will make the expense of a stone drain far exceed that of a tile drain, with tiles at fair prices. The tiles, if well secured at the inlet and outlet of the drain, will entirely exclude rats and mice, which always infest stone drains to cellars. Care must be taken, if the water is conducted on the surface of the cellar into the drain, that nothing but pure water be admitted. This may be effected by a fine strainer of wire or plate; or by a cess-pool, which is better, because it will also prevent any draft of air through the drain.

The very best method of draining a cellar is that adopted by the writer, on his own premises. It is, in fact, a mere [354] application of the ordinary principles of field drainage. The cellar was dug in sand, which rests on clay, a foot or two below the usual water-line in winter, and a drain of chestnut plank laid from the cellar to low land, some 20 rods off. Tiles were not then in use in the neighborhood, and were not thought of, when the house was built.

In the Spring, water came up in the bottom of the cellar, and ran out in little hollows made for the purpose, on the surface.

Not liking this inconvenient wetness, we next dug trenches a few inches deep, put boards at the sides to exclude the sand, and packed the trenches with small stones. This operated better, but the mice found pleasant accommodations among the stones, and sand got in and choked the passage. Lastly, tiles came to our relief, and a perfect preventive of all inconvenient moisture was found, by adopting the following plan:

The drain from the cellar was taken up, and relaid 18 inches below the cellar-bottom, at the outlet. Then a trench was cut in the cellar-bottom, two feet from the wall, a foot deep at the farthest corner from the outlet, and deepening towards it, round the whole cellar, following the course of the walls. In this trench, two-inch pipe tiles were laid, and carefully covered with tan-bark, and the trenches filled with the earth. This tile drain was connected with the outlet drain 18 inches under ground, and the earth levelled over the whole. This was done two years ago, and no drop of water has ever been visible in the cellar since it was completed. The water is caught by the drain before it rises to the surface, and conducted away.

Vegetables of all kinds are now laid in heaps on the cellar-bottom, which is just damp enough to pack solid, and preserves vegetables better, in a dry cellar, than casks, or bins with floors. [355]

A little sketch of this mode of draining cellars, representing the cellar referred to, will, perhaps, present the matter more clearly.

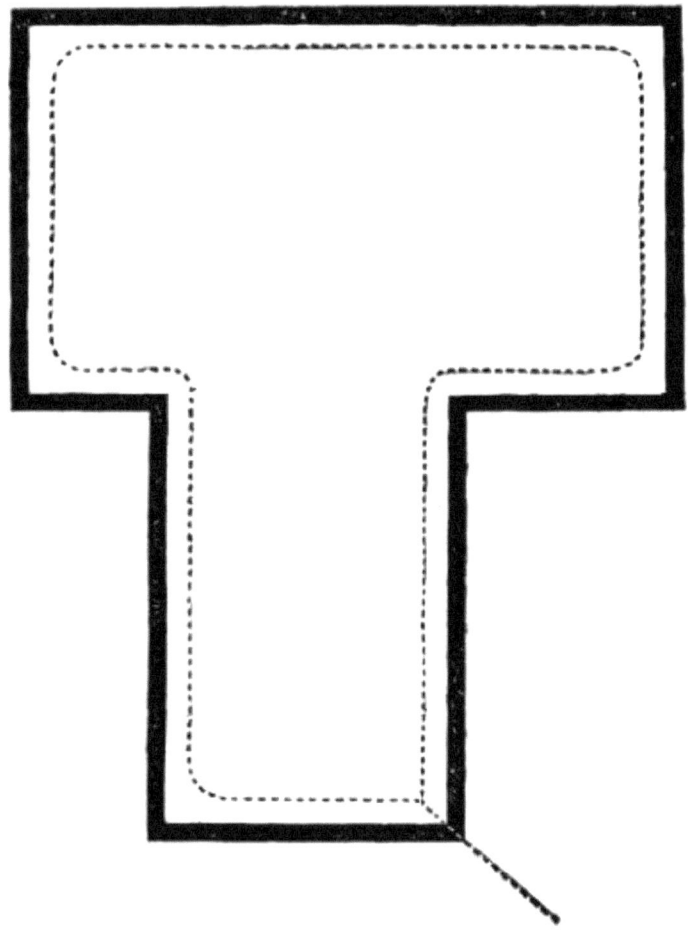

Fig. 99—Drainage of Cellar.

Many persons have attempted to exclude water from their cellars by cementing them on the bottom, and part way up on the sides. This might succeed, if the cellar wall were laid very close, and in cement, and a heavy coating of cement applied to the bottom. A moment's attention to the subject will show that it is not likely to succeed, as experience shows that it seldom, if ever, does.

The water which enters cellars, frequently runs from [356] the surface behind the cellar wall, where rats always keep open passages, and fills the ground and these passages; especially when the earth is frozen, to the surface, thus giving a column of water behind the wall six or eight feet in height. The pressure of water is always in proportion to its height or head, without reference to the extent of surface. The pressure, then, of the water against the cemented wall, would be equal to the pressure of a full mill-pond against its perpendicular dam of six or eight feet height! No sane man would think of tightening a dam, with seven feet head of water, by plastering a little cement on the down-stream side of it, which might as well be done, as to exclude water from a cellar by the process, and under the conditions, stated.

## DRAINAGE OF BARN CELLARS.

Most barns in New England are constructed with good substantial cellars, from six to nine feet deep, with solid walls of stone. They serve a three-fold purpose; of keeping manure, thrown down from the cattle and horse stalls above; of preserving turnips, mangolds, and other vegetables for the stock; and of storing carts, wagons, and other farm implements. Usually, the cellar is divided by stone, brick, or wood partitions, into apartments, devoted to each of the purposes named. The cellar for manure should not be wet enough to have water flow away from it, nor dry enough to have it leach. For the other purposes, a dry cellar is desirable.

Perhaps the details of the drainage of a barn cellar on our own premises, may give our views of the best mode of drainage, both for a manure cellar, and for a root and implement cellar. The barn was built in 1849, on a site sloping slightly to the south. In excavating for the wall, at about seven feet below the height fixed for the sills, we came upon a soft, blue clay, so nearly fluid [357] that a ten-foot pole was easily thrust down out of sight, perpendicularly, into it! Here was a dilemma! How could a heavy wall and building stand on that foundation? A skillful engineer was consulted, who had seen heavy brick blocks built in just such places, and who pronounced this a very simple case to manage. "If," said he, "the mud cannot get up, the wall resting on it cannot settle down." Upon this idea, by his advice, we laid our wall, on thick plank, on the clay, so as to get an even bearing, and drove down, against the face of the wall, edge to edge, two-inch plank to the depth of about three feet, leaving them a foot above the bottom of the wall. Against this, we rammed coarse gravel very hard, and left the bottom of the cellar one foot above the bottom of the wall, so that the weight might counterbalance the pressure of the wall and building. The building has been in constant use, and appears not to have settled a single inch.

The cellar was first used only for manure, and for keeping swine. It was quite wet, and grew more and more so every year, as the water found passages into it, till it was found that its use must be abandoned, or an amphibious race of pigs procured. It was known, that the most of the water entered at the north corner of the building, borne up by the clay which comes to within three feet of the natural surface; and, as it would be ruinous to the manure to leach it, by drawing a large quantity of water through it into drains, in the usual mode of draining, it was concluded to cut off the water on the outside of the building, and before it reached the cellar. Accordingly, a drain was started at the river, some twenty rods below, and carried up to the barn, and then eight feet deep around two sides of it, by the north corner, where most water came in.

We cut through the sand, and four or five feet into the clay, and laid one course only of two-inch pipe-tiles at the [358] bottom. As this was designed for a catch-water, and not merely to take in water at the bottom, in the usual way, we filled the trench, after covering the tiles with tan, with coarse sand above the level of the clay, and put clay upon the top. We believe no water has ever crossed this drain, which operates as perfectly as an open ditch, to catch all that flows upon it. The manure cellar was then dry enough, but the other cellar was wanted for roots and implements, and the water was constantly working up through the soft clay bottom, keeping it of the consistency of mortar, and making it difficult to haul out the manure, and everyway disagreeable.

One more effort was made to dry this part. A drain was opened from the highway, which passes the barn, to the south corner; and about two and a half feet below the bottom of the cellar, along inside the wall, at about three feet distance from it, on two of the sides; and another in the same way, across the middle of the cellar.

These, laid with two-inch tiles, and filled with gravel, were connected together, and led off to the wayside. The waste water of two watering places, one in the cellar, and another outside, supplied by an aqueduct, was conducted into the tiles, and thus quietly disposed of. The reason why the drains are filled with gravel is, that as the soft clay, in which the tiles were laid, could never have the heat of the direct rays of the sun on its surface, there might be no cracking of it, sufficient to afford passage for the water, and so this was made a catch-water to stop any water that might attempt to cross it.

The work was finished last Autumn, and we have had but the experience of a single season with it; but we are satisfied that the object is attained. The surface of the implement cellar, which before, had been always soft and muddy, has ever since been as dry and solid as a highway in Summer; and the root cellar, which has a cemented [359] bottom, is as dry as the barn floor. The manure can now be teamed out, without leaving a rut, and we are free to confess, that the effect is greater than we had deemed possible.

The following cut will show at a glance, how all the drains are laid, the dotted lines representing the tile drains:

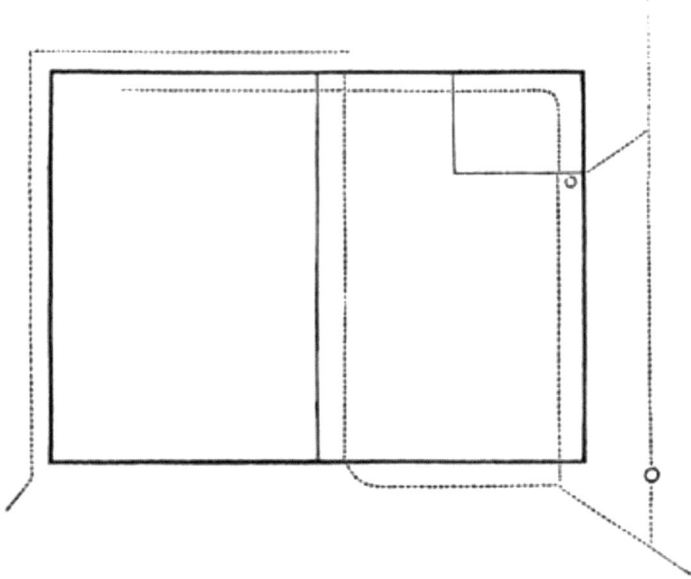

Fig. 100.

The drain outside the barn, on the right, leads from a spring, some two hundred feet off, into the cellar and into the yard, and supplies water to the cattle, at the points indicated. The waste water is then conducted into the drains, and passes off.

# CHAPTER XXIII [360]
# DRAINAGE OF SWAMPS.

Vast Extent of Swamp Lands in the United States. — Their Soil. — Sources of their Moisture. — How to Drain them. — The Soil Subsides by Draining. — Catch-water Drains. — Springs. — Mr. Ruffin's Drainage in Virginia. — Is there Danger of Over-draining?

In almost, if not quite every State, extensive tracts of swamp lands are found, not only unfit, in their natural condition, for cultivation, but, in many instances, by reason of obnoxious effluvia, arising from stagnant water, dangerous to health.

Of the vast extent of such lands, some idea may be formed, by adverting to the fact, that under the grants by Congress, of the public lands given away to the States in which they lie, as of no value to the Government and as nuisances to their neighborhood, in their natural condition; sixty millions of acres, it is estimated, will be included.

These are only the public lands, and in the new States. In every township in New England, there are hundreds of acres of swamp land, just beginning to be brought to the notice of their owners, as of sufficient value to authorize the expense of drainage.

To say that these swamps are the most fertile and the most valuable lands in New England, is but to repeat the assertion of all who have successfully tried the experiment of reclaiming them.

In their natural state, these swamps are usually covered with a heavy growth of timber; but the greater portion [361] of them have been partially cleared, and many of them are mowed, producing a coarse, wild, and nearly worthless grass.

The soil of these tracts is usually a black mud or peat, partly the product of vegetable growth and decay on the spot, and partly the deposit of the lighter portion of the upland soil, brought down by the washing of showers, and by spring freshets. The leaves of the surrounding forest, too, are naturally dropped by the Autumn winds into the lowest places, and these swamps have received them, for ages. Usually, these lands lie in basins among the hills, sometimes along the banks of streams and rivers, always at the lowest level of the country, and not, like Irish bogs, upon hill-tops, as well as elsewhere. Their surface is, usually, level and even, as compared with other lands in the old States. Their soil, or deposit, is of various depth, from one foot to twenty, and is often almost afloat with water, so as to shake under the feet, in walking over it.

The subsoil corresponds, in general, with that of the surrounding country, but is oftener of sand than clay, and not unfrequently, is of various thin strata, indicating an alluvial formation. Frogs and snakes find in these swamps an agreeable residence, and wild beasts a safe retreat from their common foe. Notoriously, such lands are unhealthful, producing fevers and agues in their neighborhood, often traceable to tracts no larger than a very few acres.

In considering how to drain such tracts, the first inquiry is as to the source of the water. What makes the land too wet? Is it the direct fall of rain upon it; the influx of water by visible streams, which have no sufficient outlet; the downflow of rain and snow water from the neighboring hills; or the bursting up of springs from below? [362]

Examine and decide, which and how many, of these four sources of moisture, contribute to flood the tract in question. We assume, that the swamp is in a basin, or, at least, is the lowest land of the neighborhood. The three or four feet of rain water annually falling upon it, unless it have an outlet, must make it a swamp, for there can usually be no natural drainage downward, because the swamp itself is the lowest spot, and no adjacent land can draw off water from its bottom. Of course, there is lower land towards the natural outlet, but usually this is narrow, and quite insufficient to allow of drainage by lateral percolation. Then, always, more or less water must run upon the surface, or just below it, from the hills, and usually, a stream is found in the swamp, if none pours into it from above.

The first step is a survey, to ascertain the fall over the whole, and the next, to provide a deep and sufficient outlet. Here, we must bear in mind a peculiarity of such lands. All land subsides, more or less, by drainage, but the soils of which we are speaking, far more than any other. Marsh and swamp lands often subside, or *settle*, one or two feet, or even more. Their soil, of fibrous roots, decayed leaves, and the like, almost floats; or, at least, expands like a sponge; and when it is compacted, by removing the water, it occupies far less space than before. This fact must be kept in mind in all the process. The outlet must be made low enough, and the drains must be made deep enough, to draw the water, after the subsidence of the soil to its lowest point.

If a natural stream flow through, or from, the tract, it will usually indicate the lowest level; and the straightening and clearing out of this natural drain, may usually be the first operation, after opening a proper outlet. Then a catch-water open drain, just at the junction of the high and low land, entirely round the swamp, will be necessary [363] to intercept the water flowing into the swamp. This water will usually be found to flow in, both on the surface, and beneath it, and in greater or less quantities, according to the formation of the adjacent land. This catch-water is essential to success. The wettest spot in a swamp is frequently, just at its edge, because there the surface-water is received, and because there too, the water that has come down on an impervious subsoil stratum, finds vent. It is in vain to attempt to lay dry a swamp, by drains, however deep, through its centre. The water has done its mischief, before it reaches the centre. It should be intercepted, before it has entered the tract, to be reclaimed.

This drain must be deep, and therefore, must be wide and sloping, so that it may be kept open; and it should be curved round, following the line of the upland to the outlet. Often it has been found, in England, that a single drain, six or eight feet deep, has completely drained a tract of twenty or thirty acres, by cutting off all the sources of the supply of water, except that from the clouds. This kind of land is very porous and permeable, and readily parts with its water, and is easily drained; so that the frequent drains necessary on uplands, are often quite unnecessary. Many instances are given, of the effect of single deep drains through such tracts, in lowering the water in wells, or entirely drying them, at considerable distances from the field of operation.

When the surface-water and shallow springs have thus been cut off, the drainer will soon be able to determine, whether he has effected a cure of his dropsical patient. Often it will be found, that deep seated springs burst up in the middle of these low tracts, furnishing good and pure water for use. These, being supplied by high and distant fountains, run under our deepest drains, and find vent through some fracture of the subsoil. They diffuse their ice-cold water through the soil, and prevent the growth [364] of all valuable vegetation. To these, we must apply Elkington's system, and hit them *right in the eye*! by running a deep drain from some side or central drain, straight to them, and drawing off the water low enough beneath the surface to prevent injury. A small covered drain with two-inch pipes, will usually be sufficient to afford an outlet to any such spring.

When we have thus disposed of the water from the surface-flow, the shallow springs and the deep springs, and given vent to the water accumulated and ponded in the low places, we have then accomplished all that is peculiar to this kind of drainage. We have still the water from the clouds, which is twice as much as will evaporate from a land-surface, to provide for. We assume that this cannot pass directly down by percolation, because the subsoil is already saturated; and therefore, even if all the other sources of wetness are cut off, we shall still have a tract of land too wet for wheat and corn. If the swamp be very small, these main ditches may sufficiently drain it; but if it be extensive, they probably will not. We have seen that we have some eighteen or twenty inches of water to be disposed of by drainage; so much that evaporation cannot remove consistently with good cultivation; and, although this amount might, in a very deep peaty soil, percolate to a great distance laterally, to find a drain, yet in shallow soil resting on a retentive subsoil, drains might be necessary at distances similar to those adopted on wet upland fields. To this part of the operation, we should, therefore, apply the ordinary principles of drainage, putting in covered drains with tiles, if possible, at four feet depth or more, ordinarily, and at distances of from forty to sixty feet, although four-foot drains at even one hundred feet distance, in peat and black mud, might often be found sufficient.

Through the kindness of Edmund Ruffin, Esq., of Virginia, [365] we have been furnished with three elaborate and valuable essays, on the drainage and treatment of flat and wet lands in lower Virginia and North Carolina, published in the Transactions of the Virginia State Agricultural Society, for 1857. The principal feature of his system is based upon his correct knowledge of the geological formation of that district; of the fact in particular, that, underlying the whole of that low country, there is a bed of pure sand lying nearly level, and filled with water, which may be drawn down by a few large deep drains, thus relieving the surface-soil of surplus water, by comprehensive but simple means.

We have before referred to Mr. Ruffin as the publisher, more than twenty years ago, of "Elkington's Theory and Practice of Draining, &c., by Johnstone;" and we find in his recent essays, evidence of how thoroughly practical he has made the system of Elkington in his own State. Indeed, we know of no other American writer who records any instance of marked success in the use of Elkington's peculiar idea of releasing pent up waters by boring. Mr. Ruffin, however, has applied, with great success, this principle of operation, to the saturated sand-beds which underlie the tracts of low land in his district of country. These water-beds in the sand lie at depths varying usually from four to eight feet below the surface. This surface stratum is comparatively compact, and very slowly pervious to water before it is drained. The water from below, is constantly pressing slowly up through it, of course preventing any downward percolation of the rain-water. By running deep drains at wide intervals, and boring down through this surface stratum with an auger, the pent up water below finds vent and gushes up in copious springs through the holes, and flows off without coming nearer to the surface than the bottom of the drains; thus relieving the pressure upward, and lowering the water-line in proportion to the depth of the drains. [366]

Mr. Ruffin gives an instance of the drying up of a well half a mile distant, by cutting a deep drain into this sand-bed, and thus lowering its water-line.

No doubt in many localities in our country, a competent geological knowledge may detect formations where this principle of drainage may be applied with perfect success, and with great economy.

*Is there danger of over-draining swamp lands?* In speaking of the injury by drainage, we have treated of this question.

Our conclusions may be briefly stated here. There is an impression among English writers, that light peaty soils may be too much drained; but many distinguished drainers doubt the proposition. No doubt there are soils too porous and light to be productive, when first drained. They may require a season or two to become compact, and may require sand, or clay, or gravel, to give them the requisite density; but these soils would, we believe, be usually unproductive if shallow drained.

In short, our idea is, that, in general, a soil so constituted as to be productive under any circumstances, will retain, by attraction, moisture enough for the crops, though intersected by four-foot drains at usual distances; and that cold water pumped up to the roots from a stagnant pool at the bottom, is not, either in nature or art, a successful method of irrigation.

Still we believe that peaty soils may be usually drained at greater distances, or by shallower drains, than most uplands, because of their more porous nature; and we should advise inexperienced persons not to proceed with a lavish expenditure of labor to put in parallel drains at short distances, till they have watched, for a season, the operation of a cheaper system. They may thus attain the desired object, with the smallest expense. If the first drains are judiciously placed, and are found insufficient, others may be laid between the first, until the drainage is complete.

# CHAPTER XXIV [367]
# AMERICAN EXPERIMENTS IN DRAINAGE—
# DRAINAGE IN IRELAND.

Statement of B. F. Nourse, of Maine.—Statement of Shedd and Edson, of Mass.—Statement of H. F. French, of New Hampshire.—Letter of Wm. Boyle, Albert Model Farm, Glasnevin, Ireland.

It was part of the original plan of this work, to give a large number of statements from American farmers of their success in drainage; but, although the instances are abundant, want of space limits us to a few. These are given with such diagrams as will not only make them intelligible, but, it is hoped, will also furnish good examples of the arrangement and modes of executing drains, and of laying them down upon plans for future reference. The mode adopted by Shedd and Edson, of indicating the size of the pipes used, by the number of dots in the lines of drains, is original and convenient. It will be seen by close attention, that a two-inch pipe is denoted by dots in pairs, a three-inch pipe by dots in threes, and so on.

It is believed that Mr. Nourse's experiment is one of the most thorough and successful works of drainage yet executed in America. His plan is upon page 195.

## STATEMENT OF B. F. NOURSE, ESQ.

Goodales Corner, Orrington, Me.,
Sept. 1st, 1858.

My dear Sir:—So much depends upon the preliminary surveys and "levels" for conducting works of thorough-draining and irrigation cheaply, yet to obtain the most beneficial results, that a competent person, [368] such as an engineer or practiced land-drainer, should be employed to make them, if one can be obtained. Unfortunately for me, when I began this operation, some years ago, there were no such skilled persons in the country, or I could learn of none professionally such, and was forced to do my own engineering. Having thus practically acquired some knowledge of it, I use and enjoy a Summer vacation from other pursuits, in the prosecution of this; and this employment, for the last few weeks, has delayed my answer to your inquiries. Nor could I sooner arrive at the figures of cost, extent, &c., of this season's work.

This is expected to be completed in ten days, and then I shall have laid, of

| | | |
|---|---|---|
| Stone drains, including mains | 702 | rods |
| Tile drains (two inches, or larger) | 1043 | " |
| In all | 1745 | " |

or, about five and one-half miles, laying dry, *satisfactorily*, about thirty-five acres. The character and extent of the work will better appear by reference to the plan of the farm which I send with this for your inspection.

The earlier portion was fairly described by the Committee of the Bangor Hort. Soc. — (See Report, for 1856, of the Maine Board of Agriculture.) It was far too costly, as usual in works of a novel character conducted without practical knowledge. No part of my draining, even that of this season, has been done so cheaply as it ought to be done in Maine, and will be done when tiles can be bought at fair prices near at hand. I call your attention particularly to this, because the magnitude of the cost, as I represent it, ought not to be taken as a necessary average, or standard outlay per acre, by any one contemplating similar improvement, when almost any farmer can accomplish it equally well at far less cost. My unnecessary expenditures will not have been in vain, if they serve as a finger-post to point others in a profitable way.

My land had upon its surface, and mingled in its super soil, a large quantity of stones, various in size, from the huge boulders, requiring several blasts of powder to reduce them to movable size, to the rubble stones which were shoveled from the cart into the drains. To make clean fields all these had to be removed, besides the many "heaps" which had been accumulated by the industry of my predecessors. A tile-drain needs no addition of stone above the pipe; indeed, the stone may be a positive injury, as harboring field vermin, or, if allowed to come within two feet of the surface, as obstructing deep tillage, and favoring the access of particles of soil upon or into the tile with the [369] rapid access of water which they promote. Carefully placed to the depth of six or eight inches in a four-foot drain, quite small stones are, perhaps, useful, and they certainly facilitate the drawing of water from the surface. Such was, and still is, with many, the prescribed method of best drainage in Scotland, and some parts of England. The increased cost of adding the stone above the tile is obvious; and when the width of that drain is enlarged to receive them, the cost is materially enhanced. Yet such has been my practice, at first, under the impression of its necessity, and all the time from a desire to put to use, and out of sight, the small stones with which I was favored in such abundance. The entire cost of moving, and bringing more than 2,500 heavy loads of stone, is included in the cost of drains, as set down for the 1,745 rods.

Including this part of expense, which is never *necessary* with tile, and cannot be incurred in plain clay soils, or clay loams free of stones, the last 700 rods cost an average of 97 cents per rod completed. This includes the largest mains; of which, one of 73 rods was opened four feet wide at bottom of the trench, of which the channel capacity is $18 \times 18 = 324$ square inches, and others 110 rods of three and one-half and three feet width at bottom, all these mains being laid entirely with stone. The remainder of the 700 rods was laid with two-inch tile, which cost at the farm eighteen dollars per 1,000. These last were opened four rods apart, and lay dry about seventeen acres, at a cost, including the mains, of $678, or $40 per acre. In this is included every day's labor of man and beast, and all the incidental expenses, nothing being contributed by the farm, which is under lease.

I infer that an intelligent farmer, beginning aright, and availing himself of the use of team and farm labor, when they can best be spared from other work—as in the dry season, after haying—or paying fair prices for digging his ditches only, and doing the rest of the work from the farm, can drain thoroughly at a cost of $20 per acre, drains four rods apart, and four feet deep; or at $25 per acre, forty feet apart, and three feet nine inches deep.

My subsoil is very hard, requiring constant use of the pick, and sharpening of the picks every day, so that the labor of loosening the earth was one-third or one-half more than the throwing out with a shovel. The price paid per rod, for opening only, to the depth of three and a half feet (or, perhaps, three and three-quarters average,) of a width for laying tile, was 25 cents per rod. At this price, the industrious men, skillful with tools, earned $1.12 to $1.25 per day, besides [370] board; and they threw out one-third more earth than was really necessary, for "room to work" as they said. *But they labored hard, 14 hours per day.* The same men, working in a soil free from stones, and an easier subsoil, would, in the same time, open from 50 to 100 per cent. more length of ditch.

The greater part of these drains were laid four rods apart. When first trying this distance upon a field, of which the soil was called "springy and cold," and was always too wet in the Spring and early Summer for plowing, a partial, rather than "thorough" drainage was attempted, with the design, at some future day, to lay intermediate drains. The execution of that design may yet appear expedient, although the condition of soil already obtained, is satisfactory beyond expectation.

Owing to the excess of water that saturated the soil in Spring and Fall, the former proprietors of the farm had not attempted the cultivation of the field alluded to, for many years. Originally producing heavy crops of hay, it had been mowed for thirty years or more, and was a good specimen of "exhausted land," yielding one-half or three-fourths of a ton of hay per acre. This field is designated in the plan, as the "barley field, 1858," lies south-west of the dwelling-house, and contains nearly six acres. Its northerly half, being the lower end of the field, was drained in 1855, having been Summer-plowed, and sowed with buckwheat, which was turned under, when in flower, as a fallow crop. The other half was drained in 1856; plowed and subsoiled the same Fall. In 1857, nearly the whole field was planted with roots—potatoes, rutabagas, mangolds, carrots, English turnips, &c.—and one acre in corn. For these crops, fair dressings of manure were applied—say ten or twelve cartloads of barn-manure plowed in, and one hundred pounds of either guano or bone-dust harrowed in, or strewed in the drill, for each acre; about fifteen loads per acre of seasoned muck or peat were also plowed in. There was a good yield of all the roots; for the corn, the season was unfavorable. Last Spring, a light dressing of manure, but all that we could afford, was applied, the whole well ploughed, harrowed, seeded to grass with barley, harrowed, and rolled. The barley was taken off last week; and, from the five and three-quarter acres, seventeen heavy loads were hauled into the barn, each estimated to exceed a ton in weight. The grain from a measured acre was put apart to be separately threshed, and I will advise of its yield when ascertained. [A] This was said, by the many farmers who saw it, [371] including some from the Western States, to be the "handsomest field of grain" they had ever seen. The young grass looks well; and I hope, next Summer, to report a good cut of "hay from drained land."

Last Winter, there were no snows to cover the ground for sleighing until March; and, lying uncovered, our fields were all frozen to an unusual depth. But, *our drains did not cease to run through the Winter.* And Mr. O. W. Straw, who works the farm, and was requested to note the facts accurately, wrote to me this Spring, "the frost came out of the drained land about one week first" (that is, earlier than from the undrained land adjacent); and, "in regard to working condition, the drained land was in advance of the undrained, ten days, at least." The absence of snow permitting this unusual depth of frost, had caused a rare equality of condition the last Spring, because, until the frost was out, the drains would not draw surface-water. Usually, when early snows have fallen to protect the ground, and it remains covered through the Winter, the frost goes off with the snow, *or earlier*, and, within a few days, the land becomes in good condition for plowing—quite two weeks earlier than the driest of my undrained fields, or any others in the vicinity.

These remarks apply to land in which the drains are four rods apart. The farm lies with an inclination northerly and easterly, the fall varying from 1 in 33 to 1 in 8; that in most of the drains laid four rods apart, being about 1 in 25. The drains in the "barley field" fall 1 in 27, average, all affording a rapid run of water, which, from the mode of construction, and subsequent subsoiling, finds ready access to the drain-channels. Hence, we never observe running water upon the surface of any of our drained lands, either during the heaviest rains, or when snows are melting, and the wasteful "washing" from the surface that formerly injured our plowed grounds, has ceased.

It is fair to suppose that it is the considerable descent which renders the drains so effectual at four rods apart; and that where there is but slight fall, other circumstances being the same, it would be necessary to lay drains much nearer, for equal service.

The results of one man's experiments, or practice, whether of success or failure, should not be conclusive to another, unless all the circumstances are identical. These are ever varying from one farm to another; and only a right understanding of the natural laws or principles brought into use, can determine what is best in each case. Therefore, a description of the methods I have used, or any detailed suggestions I may give, as the result of experience, would not be worth much, unless tested by the well-ascertained rules applicable to them, which men of [372] science and skill have adopted and proved, by the immensely extended draining operations in Great Britain, and those begun in this country. These are now given in elaborate treatises, and quoted in agricultural journals. But they should be made familiar to every farmer, in all their practical details, and with methods suited to our country, where labor is dear and land cheap, as contrasted with the reversed conditions in England, where the practice of "thorough-draining" has so generally obtained, and has so largely improved the conditions of both landlord and tenant. Your book will do this, and thus do a great good; for draining will greatly enlarge the productive capacity of our land, and, consequently, its value, while it will render labor more effective and more remunerative to the employer and the employed.

The fact of increased production from a given quantity of land, by draining, being ascertained beyond question, and the measure of that increase, at its minimum, being more than the interest at six per cent. upon the sum required to effect it—even at $50 per acre—the question of expediency is answered. To the owner of tillage lands there is no other such safe, sure, and profitable investment for his money. He lodges it in a bank that will never suspend payments, and from which better than six per cent. dividend can be received annually.

[A] This was threshed about the middle of November, and yielded "51 bushels, round measure." The entire field averaged 45 bushels per acre.

## STATEMENT OF SHEDD AND EDSON.

Boston, February 1, 1859.

Dear Sir:—The plan for a system of thorough drainage, a copy of which we send you herewith, was executed for Mr. I. P. Rand, of Roxbury.

An outfall was obtained, at the expense of considerable labor, by deepening the Roxbury and Dorchester Brook for a distance of nearly a quarter of a mile, about four hundred feet of which was through a rocky bottom, which required some blasting. The fall thus obtained was only about two inches in the whole distance.

The fall which can be obtained for the main drain is less than two inches per hundred feet, but the lateral drains entering into the main, will have a fall varying from two inches to a foot per hundred.

The contour lines, or lines traced along the ground, intersecting points on an equal level, are drawn on this plan, showing a fall of four-tenths of a foot, each line being in every part four-tenths of a foot lower than the line above it. Where the lines are near together, the fall is greater, as a less horizontal distance is passed over before reaching a point which is four-tenths lower than the line above.

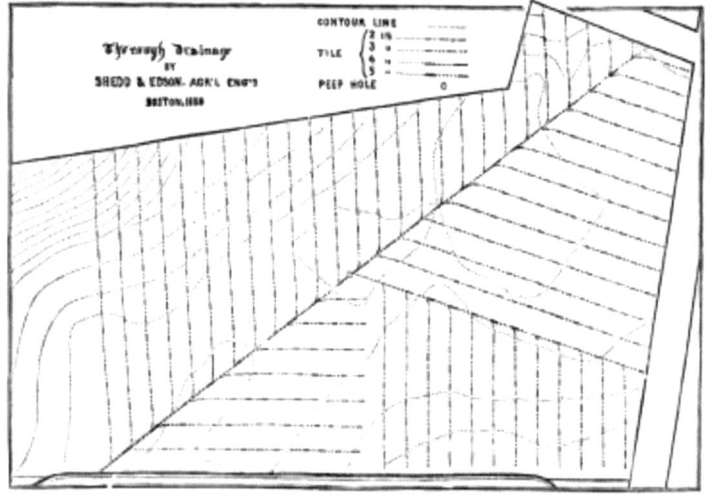

[373] It will be seen by the plan, that the fall in the line occupied by the main drain is very slight, while the side drains have a fall much greater.

The lateral drains are run in the line of steepest descent, which is, of course, at right angles to the general direction of the contour lines.

The water from the entire system is collected, and escapes at one outlet into the brook.

A peep hole is placed at the intersection of the sub-main drain with the main, which commands about one-half the entire area — the other, half is commanded by the outlet.

Two-inch tile will be laid in the lateral drains, and three, four, and five-inch in the sub-main and main.

It is quite indispensable, to the successful execution of a plan of drainage on land so level as this, that careful measurements be made on the ground with an engineer's level, and such a representation of its surface projected as will show to the eye at a glance what all the natural inclinations are. The work can then be laid out with ease in the best position, and executed in a systematic manner. The time and labor which is devoted to such an examination of the ground is well spent, and, with the knowledge gained by it, the work can be carried on with such economy as to save the original cost of the examination many times over.

Very truly, yours, Shedd & Edson
Hon. H. F. French, Exeter, N. H.

## STATEMENT OF HENRY F. FRENCH, OF EXETER, N. H.

The drained field represented in the plan (Fig. 102), contains about eight acres. I purchased it in 1846. The upper part of it is sand, with underlying clay at depths of from four to ten feet. The field slopes towards the river, and, on the slope, the clay strata coming out to the surface, naturally bring out the water, so that the side hill was so wet as to produce cranberries—quite too wet for any hoed crop. At the foot of the hill the soil is a stiff clay, with veins of sand and gravel. Through the centre was a wet ravine, which served as a natural outlet for the springs, and which was so full of black alders as to make an excellent cover for woodcock. Until the land was drained, this ravine was impassable in the hay season even, except by a bridge which I built across it. Now it may be crossed at any season and at any point.

I first attempted to drain the wettest parts with brush drains, running them into the wet places merely, and succeeded in drying the land sufficiently to afford good crops of hay. I laid one brush-drain across [374] the brow of the hill, five feet deep, hoping to cut off all the water, which I supposed ran along upon the surface of the clay. This dried the land for a few rods, but the water still ruined the lower parts of the field, and the drain produced very little effect upon the land above it. In 1856, finding my brush drains quite in-sufficient, I thorough-drained the side-hill on the lower part of the plan at the reader's left hand, at fifty feet distances, up and down the slope, at an average of about four feet depth, going five feet deep on the brow of the hill, to cut through the brush-drain. I used two-inch sole-tiles for minors, and three-inch for the main.

The effect was instantaneous. The land which, in the Spring of 1856, had been so wet that it could not, even though partially drained with brush-drains, be planted till the 5th of June, was, in

1857, ready to work as soon as the snow was off. My farm journal says, under date of April 6th, "plowed drained land with double plow two days after a heavy storm—dry enough." I spent that Summer in Europe. The land was planted with corn, which produced a heavy crop. I find an entry in my journal, on my return, "My drained land has been in good condition—neither too wet nor too dry—all Summer."

In the Fall of 1857, I laid about 170 rods in other parts of the field, at similar depths and distances, and in 1858 completed the upper part, on which is an orchard of apple trees. A part of this orchard was originally so wet as to kill the trees the first year, but by brush-drains I dried it enough to keep the next set alive. There was no water visible at the surface, and the land was dry enough for corn and potatoes; still the trees looked badly, and many were winter-killed. I had learned the formation of the earth about my premises, of which I had at first no adequate conception, and was satisfied that no fruit tree could flourish with its feet in cold water, even in Winter. All nursery-men and fruit-growers agree, that land must be well drained for fruit. I therefore laid four-foot tile drains between the rows of trees, in this apparently dry sand. We found abundance of water, in the driest season, at four feet, and it has never ceased to flow copiously.

I measured accurately the discharge of water from the main which receives the drainage of about one and a half acres of the orchard, at a time when it gave, what seemed to me an average quantity for the Winter months, when the earth was frozen solid, and found it to be about 480 barrels per day! The estimate was made by holding a bucket, which contained ten quarts, under the outlet, when it was found that it would fill in fifteen seconds, equal to ten gallons per minute; and six hundred gallons, or twenty barrels per hour, and four hundred and eighty barrels per day. [375]

I have seen the same drain discharge at least four times that quantity, at some times! The peep-holes give opportunity for inspection, and I find the result to be, that the water-table is kept down four feet below the surface at all times, except for a day or two after severe rain-storms.

There is an apparent want of system in this plan, partly to be attributed to my desire to conform somewhat to the line of the fences, and partly to the conformation of the land, which is quite uneven. At several points near the ravine, springs broke out, apparently from deep fountains, and short drains were run into them, to keep them below the surface.

The general result has been, to convert wet land into early warm soil, fit for a garden, to render my place more dry and healthful, and to illustrate for the good of the community the entire efficiency of tile-drainage. The cost of this work throughout, I estimate at fifty cents per rod, reckoning labor at $1 per day, and tiles at $12 per thousand, and all the work by hand-tools. I think in a few years, we may do the same work at one-half this cost. Further views on this point are given in the chapter on the "Cost of Drainage."

After our work was in press, we received from Mr. William Boyle, Farmer at the Albert Model Farm in Ireland, the paper which is given below, kindly sent in reply to a series of questions proposed by the author. The Albert Model Farm is one of the Government institutions for the promotion of agriculture, by the education of young men in the science and the practice of farming; and from what was apparent, by a single day's examination of the establishment in our visit to it in August, 1857, we are satisfied of its entire success. The crops then growing were equal, if not superior, to any we have seen in any country. Much of the land covered by those crops is drained land; and having confidence that the true principles of drainage for that country must be taught and practiced at this institution, we thought it might be instructive, as well as interesting to the farmers of America, to give them the means of comparison between the system there approved, and those others which we have described. [376]

Had the paper been sooner received, we should have referred to it earlier in our book; yet coming as it does, after our work was mostly in type, we confess to some feeling of satisfaction, at the substantial coincidence of views entertained at the Albert Model Farm, with our own humble teachings. With many thanks to Mr. Boyle for his valuable letter, which we commend to our readers as a reliable exposition of the most approved principles of land-draining for Ireland, we give the paper entire:

Albert Model Farm, Glasnevin, Dublin,<br>
January 31, 1859.

To the Hon. Henry F. French, Exeter, N. H.:

Sir:—Your queries on land-drainage have been too long unanswered. I have now great pleasure in sending you, herewith, my views on the points noted. * * *

Pray excuse me for the delay in writing. I am, sir,
Your obliged and obedient servant, William Boyle.

## LAND DRAINAGE—REPLIES TO QUERIES, ETC.

*Introductory observations.* Ireland contains close on to twenty-one millions of acres, thirteen and a half millions of which were returned as "arable land," in 1841. By "Arterial" and thorough-drainage, &c., effected through loans granted by government, the extent of arable land has been increased to fifteen and a half millions of acres. The "Board of Works" has the management of the funds granted for drainage and land improvements generally, and competent inspectors are appointed to see that the works are properly executed. The proprietor, or farmer, who obtains a loan may, if competent, claim and obtain the appointment of overseer on his own property, and thus have an opportunity of economically expending the sum which he will have to repay (principal and interest) by twenty-two installments. The average cost of thorough-drainage, under the Board of Works, has been about £5 per statute acre. In 1847, when government granted the first loan for land-drainage, tiles were not so easily obtained as at present, nor was tile-drainage well understood in this country; and the greater part of the drains then made—and for some years after—were either sewered with stones, formed into a conduit of various dimensions, and covered over with finely-broken stones, or the latter were filled into the bottom of the drain, to about one foot in depth, as recommended by Smith, of [377] Deanston. The dimensions for minor drains, sewered with stones, were, usually, three and a half feet deep, fifteen inches wide at top, and three to four inches wide at bottom (distance apart being twenty-one feet); and the overseer carried about with him a wooden gauge, of a size to correspond, so that the workmen could see at a glance what they had to do. These drains are reported to have given general satisfaction; and they were cheaply made, as the stones were to be had in great abundance in almost every field. On *new* land, trenching was sometimes carried on simultaneously with the drainage; and it very often happened that the removal of the stones thus brought to the surface, was very expensive; but they were turned to profitable account in sewering drains and building substantial fences. In almost every case the drains were made in the direction of the greatest inclination, or fall of the land; and this is the practice followed throughout the country. Some exceptions occur on *hill-sides*, where I have seen the drains laid off at an acute angle with the line of inclination. It is not necessary that I should explain the scientific reasons for draining in the direction of the fall of the land, as that point has been fully treated of, and well illustrated, in your article already referred to. I shall now pass on to the Queries.

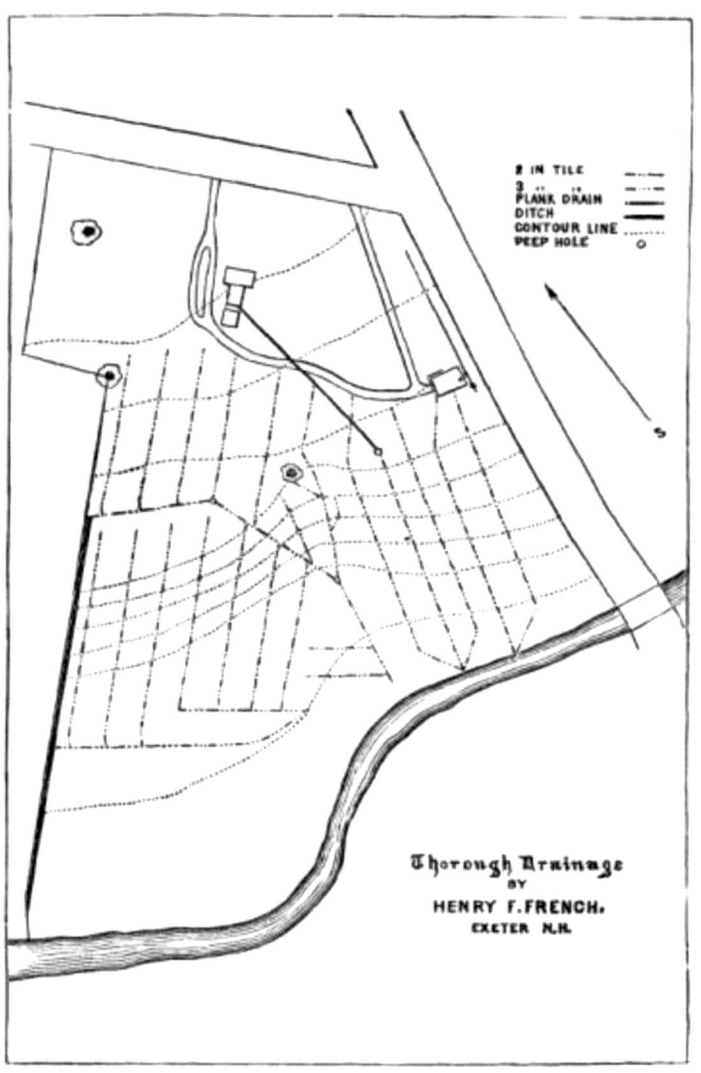

Depth of drains, and distance apart. There is still a great diversity of opinion on these points, and particularly in reference to the drainage of stiff clay soils; some of the most intelligent and practical farmers in this country hold to the opinion that, on such soils, the maximum depth should not exceed three feet, and the distance apart sixteen to twenty feet. On clay loams, having a subsoil more or less free, the general practice is, to make the drains three and a half to four feet deep, and at twenty-one to thirty feet apart. On lighter soils, having a free subsoil, four feet deep and forty feet apart are the usual limitations. This farm may be taken as a fair average of the land in Ireland, as a test for drainage; the soil is a deep clay loam; the subsoil a compact mixture of strong clay and calcareous gravel, almost free from stones. Thirty miles of drains have been made on the farm, the least distance apart being twenty-one feet, and the greatest distance thirty feet; the depth in every case, three and a half to four feet for minor drains. This drainage has given the greatest satisfaction; for although the greatest part of the work was performed by the Agricultural pupils, in training here, we have not had occasion to re-make a single drain, except in one instance, where the tiles got choked, and which I shall explain hereafter.

*Tiles: Size, Shape, Draining, Capacity, &c.* We use circular pipe tiles, of inch and a half bore, for all parallel drains whose length does [378] not exceed one hundred yards, and two-inch pipes for any additional length up to one hundred and fifty yards, the greatest length, in my opinion, a parallel drain should reach before discharging into a main or sub-main drain. We do not find it necessary to use collars on this farm, as we have *firm* ground to place the tiles on, and we can cut the drain to fit the tiles exactly. As regards the size of tiles for main and sub-main drains, *that* can only be regulated by the person in charge of the drainage at any particular place, after seeing the land opened up and the minor drains discharging. As a general rule, a circular pipe of three inches internal diameter will discharge the *ordinary* drainage of five or six statute acres, and give sufficient space for the circulation of air. It should be observed, however, that this applies to a district where the annual rain-fall is from twenty-six to thirty inches, that of all Ireland being about thirty-five inches; besides, we have not the immense falls of rain in a *few hours* that occur in other countries. All these points should be carefully considered in estimating the water-way for drainage. I have said that collars are not used with the tiles on this farm, as the bottom of the drains is quite firm and even; but, where the bed for

409

the tile is soft, and the subsoil is of a *shifting* nature, then collars should be used in every case. Collars cost about half the price of tiles, which they are made to connect, so that the use of them adds one-third to the expense of the sewering material; and, as I have already pointed out, I think it quite unnecessary to use them where the subsoil is *firm*, and where the drain can be bottomed to *fit the tile*. Where large pipes are not to be had conveniently for sewering main or sub-drains, I find a proportional number of pipes of lesser diameter to answer perfectly. It is very desirable to provide *branch pipes* for connecting the minor with the main drains. The branch should be socketed to receive the end of the last tile in the minor drain, and the point of attachment to the main pipe may be on the top or on the side of the latter. If the branch be made to lead the water into the side of the main pipe, then it should join the latter at an acute angle, that both streams may meet with the least possible opposition of forces.

*Fall necessary in Tile Drainage.* I consider one foot in one hundred yards the *least* fall to work upon with safety.

*Securing Outlets.* All the outlets from main-drains should be well secured against the intrusion of vermin, by a wrought-iron grating, built in mason-work. The water may flow into a stone trough provided with an overflow-pipe, by which the quantity discharged may be ascertained at any time, so as to compare the drainage before and after rain, [379] &c.

*Traps, or Silt Ponds.* Where extensive drainage is carried on in low-lying districts, and the principal outlet at a considerable distance, it may be found necessary to have traps at several points where the silt from the tiles will be kept. These traps may be of cast-iron, or mason-work, cemented; and provision should be made, by which they can be cleaned out and examined regularly—the drainage at these periods also undergoing inspection at the different traps.

*Plow-Draining.* We have no draining-plows in use in Ireland, that I know of; the common plow is sometimes used for marking off the drains, cutting the sides, and throwing out the earth to a considerable depth, thereby lessening the manual labor considerably. Efforts have been made in England to produce an efficient implement of this description; but it would appear there is ample room for an inventive Jonathan to walk in for a profitable patent in this department, and thus add another to the many valuable ones brought out in your great country.

*Case of Obstruction in Tiles.* Some years since, one of the principal main-drains on this farm was observed not discharging the water freely, as it hitherto had done, after a heavy fall of rain; and the land adjoining it showed unmistakable signs of wetness. The drain was opened, and traced to the point of obstruction, which was found to be convenient to a *small poplar tree*, the rootlets of which made their way into the tiles, at the depth of five and a half feet, and completely filled them, in the direction of the stream, for several yards. We have some of the tiles (horse-shoe) in our museum here, as they were then lifted from the drain, showing clearly the formidable nature of the obstruction. Another serious case of obstruction has come to my knowledge, occasioned by frogs or toads getting into the tiles of the main-drain in large numbers, on account of the outlet being insufficiently protected. In this case, a large expenditure had to be incurred, to repair the damage done.

I have not observed any case of obstruction from the roots of our cultivated plants. It has been said by some that the rootlets of mangold will reach the drains under them; and, particularly, where the drains contain most water in rapid motion. I took up the tiles from a drain on this farm, in '54, which had been laid down (by a former occupier), about the year '44, at a depth not exceeding two-and-a-half feet, and not one of these was obstructed in the least degree, although parsnips, carrots, cabbages, mangolds, &c., had been grown on this field. Obstructions may occur through the agency of *mineral* springs, [380] but very few cases of this nature are met with, at least in this country. I would anticipate this class of obstruction, if from the nature of the land there was reason to expect it, by increasing the fall in the drains and having *traps* more frequent, where the main outlets are at a distance to render them necessary. In my opinion, the roots of trees are the great intruders to be guarded against, and more particularly the *soft*-wooded sorts, such as poplars, willows, alders, &c. The distance of a drain from a tree ought always to be equal to the height of the latter.

*Tiles flattening in the drying process.* With this subject, I am not practically familiar. In most tile-works, the tiles, after passing through the moulding-machine, are placed horizontally on shelves, which rise one above another to any convenient height, on which the tiles are dried by means of heated flues which traverse the sheds where the work is carried on; or they are allowed to dry without artificial heat. I prefer the tiles prepared by the latter method, as, if sufficient time be given them to be well dried, they will burn more equally, and be more durable. The tiles will flatten more or less for the first day or two on the shelves, after which they are *rolled*. This is done by boys (who are provided with pieces of wood of a diameter equal to the bore of the tile when made), who very soon learn to get over a large number daily. The "roller" should have a shouldered handle attached, the whole thickness of which should not be greater than that of the tile. The *shoulder* is necessary to make the *ends* of the tiles even, that there may be no *very open* joints when they are placed in a drain. Once rolled, the tiles are not likely to flatten again, if the operation be performed at the proper time.

As good tiles as I ever saw were dried in a different way, and not rolled at all. As they were taken from the machine—six at a time—each carrier passed off with his tray, and placed them *on end* carefully, upon an *even floor*. When five or six rows of tiles were thus placed, the whole length of the drying-house, a board was set on edge to keep them from falling to one side; then followed five or six other rows of tiles, and so on, till the drying-ground was filled.

This was the plan adopted in a tilery near Dublin, some years ago. It is only a few days since I examined some of the tiles made at these works, which had been taken from a drain, where they had been in use for nine years; and the *clear ringing sound* produced by striking them against each other, showed what little effect that length of time produced upon them, and how well they had been manufactured. [381]

*Cost of Tiles.* We have recently paid at the works—

| For | 1½ | inch pipes | 17s. 6d. | per thousand. |
|---|---|---|---|---|
| " | 2 | " | 25s. | " |

Each tile one foot in length, and the one and one-half-inch pipes weighing 16 cwt. per thousand.

One of the great difficulties in connection with tile-making is, in many districts, to procure clay sufficiently free from lime. Tiles are very often sold by sample, sent a considerable distance, and it becomes necessary to test them, which we do (for lime) by placing them in water for a night; and, if lime is present in the tile, it will, of course, swell out, and break the latter, or leave it in a riddled state.

I have now endeavored to answer the queries in your postscript, and I have carefully avoided enlarging on some points in them with which your readers are already familiar. If I shall have thrown a single ray of additional light on this subject across the Atlantic, I shall be amply repaid for any attention I have given to thorough-drainage during the past twelve years.

I should here observe that I mislaid amongst my papers the portion of your letter containing the queries (it was a separate sheet), and it has not as yet turned up, so that I had to depend on a rather treacherous memory to keep the queries in my mind's eye. It is highly probable, therefore, that I have overlooked some of them. This circumstance was the chief cause of the delay in writing.

You are quite at liberty to make any use you please of this communication.

William Boyle.